MW01365750

Intuitive CMOS Electronics

McGraw-Hill
Series in Intuitive IC Electronics

This series will help the reader gain an intuitive understanding of electronics and computers. Mathematics is kept to a minimum, as the reader gets "inside" the devices and circuits to grasp, from the electron level up, the workings of integrated circuits, digital computers, operational amplifiers, and other electronics-related topics. The following volumes are planned for the series, and each one is written by Thomas M. Frederiksen, whose Intuitive IC Electronics *(McGraw-Hill, 1982) has proved popular with engineers, managers, students, and hobbyists worldwide.*

INTUITIVE DIGITAL COMPUTER BASICS: An Introduction to the Digital World (1988)

INTUITIVE ANALOG ELECTRONICS: From Electron to Op Amp (1988)

INTUITIVE OPERATIONAL AMPLIFIERS: From Basics to Useful Applications (1988)

INTUITIVE CMOS ELECTRONICS: The Revolution in VLSI, Processing, Packaging, and Design (1989)

INTUITIVE IC ELECTRONICS, Second Edition (1989)

Intuitive CMOS Electronics

The Revolution in VLSI, Processing, Packaging, and Design

Revised Edition

Thomas M. Frederiksen

McGraw-Hill Publishing Company
New York St. Louis San Francisco Auckland
Bogotá Hamburg London Madrid Mexico
Milan Montreal New Delhi Panama
Paris São Paulo Singapore
Sydney Tokyo Toronto

Library of Congress Cataloging-in-Publication Data

Frederiksen, Thomas M.
Intuitive CMOS electronics.

Rev. ed. of: Intuitive IC CMOS evolution. c1984.
Bibliography: p.
Includes index.
1. Integrated circuits—Very large scale integration.
2. Metal oxide semiconductors, Complementary.
I. Frederiksen, Thomas M. Intuitive IC CMOS evolution.
II. Title.
TK7874.F678 1989 621.381'73 88-22999
ISBN 0-07-021970-2
ISBN 0-07-021971-0 (pbk.)

1234567890 DOC/DOC 8921098

ISBN 0-07-021970-2

ISBN 0-07-021971-0 {PBK.}

The editors for this book were Daniel A. Gonneau and Nancy Young, the designer was Naomi Auerbach, and the production supervisor was Suzanne W. Babeuf. It was set in Century Schoolbook.

Printed and Bound by R. R. Donnelley & Sons Company

This book is dedicated to the many system engineers who, during a nationwide seminar series in 1982, found my 30-minute presentation titled "A Quick Look at the Evolution of IC Processes," interesting.

Contents

Foreword

The times when information is most valuable are during change or transition to support decision making where the facts are not clear. This book fills that need at one of the most significant transitions in semiconductor technology since the change from germanium to silicon.

Since their introduction, integrated circuits have continuously increased in density because of the significant improvements in photolithography, the reduction in defect densities, and the development of new circuit topologies. It has been a period of rapid advancement in the performance and integration of the products built with semiconductors. Without a dramatic change, however, that progress might have been limited by power density. That is, the integrated circuit density would be limited by the power that the package could effectively remove—not circuit complexity.

Fortunately, technology in the form of silicon gate oxide isolated CMOS has been developed just in time to allow the improvements in density to continue.

This book is about that transition, what it means and how it's taking place. It chronicles an industry during dramatic change and provides a clear explanation of the past, the present, and the implications for the future.

It can be a vital aid to engineers and industry observers who need to plan for the future before it arrives or for anyone who needs to stay abreast of the semiconductor industry move to CMOS.

Mark Levi
Formerly Director of CMOS Marketing
National Semiconductor Corp.
Santa Clara, California

Preface to the Revised Edition

This revised edition has incorporated much new material, and also many changes and additions were made in the sections on the capabilities and limitations of silicon wafer processing. The semiconductor industry remains very dynamic, and processing innovations are occurring very rapidly. It is impossible for any book to keep up to date with this rapid rate of change, but if the basic ideas associated with processing are understood, along with some of the technology reported by the researchers, the reader will be in a position to more rapidly understand the continuing evolution that takes place in the production lines.

An introduction to "band-gap" voltage references and how the difference between the base-emitter voltages of two matched transistors, the ΔV_{BE} voltage, is used to provide a zero temperature coefficient reference voltage has been added as background for a new CMOS op amp that has been included.

A section on Dynamic Safe Area Protection has been included because this novel technique has allowed the output power capability of ICs to be extended to 150 watts. In addition, the incorporation of large-area DMOS power transistors in power MOS ICs is also mentioned.

The story of the search for a good PNP by the IC designers has been added and the recent "vertically integrated PNP" process is described. Although many processing steps are added, this is still reported to be lower cost than dielectric isolation, a competing technology that also provides a good, compatible PNP.

A recent discovery of a way to provide an adjustable resistor, the "trimistor," by making use of the same physical mechanisms that take place in the "zener zapping" trim technology has been included. The group of engineers who worked on this project received the "Best Paper Award" at the 1987 Solid-State Circuits Conference. The physical mechanisms of this new trimistor are described and the way it was incorporated as an in-package adjustable resistor that allowed the benefits of JFET transistors in front of a bipolar op amp, without having to suffer from a large offset voltage, is covered.

The latest in high performance, CMOS op amps, has been added. A new combination of both circuit design innovations and novel device structures (including the clever use of a parasitic lateral NPN bipolar transistor in a ΔV_{BE} biasing circuit) allows these op amps to be fabricated with a standard digital CMOS process to insure compatibility with digital circuits. This eases the problem of mixing digital circuitry with high performance linear circuitry on the same chip. A novel "checkerboard-interconnected hexadecimal" (16 individual P-channel transistors), is used to provide two groups of eight paralleled transistors, each of which is used for the pair of differential input transistors of the op amp. This trick, plus careful circuit design, essentially solves the usual problem of high offset voltage in CMOS op amps. These general purpose op amps challenge the old position of supremacy that bipolar op amps have long held.

It is my hope that this book continues to provide a good intuitive introduction to wafer processing and also the background that is needed to more fully appreciate the VLSI era, and the future of ICs.

Thomas M. Frederiksen

Preface to the First Edition

Another major electronic revolution is here—the Very Large Scale Integrated (VLSI) Circuit Era. The coming changes will be more drastic than those associated with the last: The Microprocessor Revolution. These new changes will affect all of the electronic system companies, the integrated circuit (IC) designers, the IC companies, and even the system designers. In fact, the IC users of today will become the VLSI chip designers of the future.

The purpose of this book is to trace the historical evolution that has brought in the present CMOS VLSI Era. To lend perspective, Chapter 1 quickly reviews the advances that have been made in solid-state electronics and traces the evolution of both amplifying devices and logic circuits. The chapter ends with a discussion of some of the recent enhancements that have been made to the bipolar integrated circuits.

Chapter 2 traces the difficult changes that have been made in IC processes that have allowed the high volume production of Metal Oxide Semiconductor (MOS) products. This evolution is traced from the early P-channel MOS (PMOS) days, through the fast-paced, N-channel MOS (NMOS) breakthroughs, to the introduction of the early metal-gate Complementary MOS (CMOS) processes.

Chapter 3 traces the many developments that have improved the ability to fabricate complex silicon wafers. Such topics as the move from diffusion furnaces to ion implanters; the oxide isolation techniques that are both in production and are under development in research labs; the important dry etching techniques; the progress in high resolution lithography that is rapidly approaching the capability to define lines less than 1 μm in IC production; and the complex multilayered IC chip interconnect schemes.

The reasons for the present emphasis on CMOS processes for the emerging VLSI Era are discussed in Chapter 4. Many aspects of this move to CMOS VLSI are considered from the standpoint of the IC suppliers, the IC customers, and the effects on the people of both groups. An interesting way to comprehend the complexities of modern ICs is provided. The approach makes use of the more familiar ¼ inch as a scaled-up minimum resolvable dimension. Then we show the scaled-up areas that result when this scaling is applied to modern IC chips.

The wide variety of standard ICs that have been made available with the advanced CMOS processes are described in Chapter 5. These range from CMOS operational amplifiers (op amps), switched-capacitor filters and the digital-to-analog converters (DACs) and analog-to-digital (A/D) converters of linear CMOS to the many new products that are now available in digital CMOS. Changes in the telephone system that were greatly helped with CMOS products are also included.

Performance improvements of a new CMOS logic family are described and the benefits of CMOS for microprocessors and memory products are discussed. This chapter ends with a step-by-step description of a CMOS process: a multiple-layer-metal, silicon-gate wafer fabrication sequence.

Semicustom and custom circuits are discussed in Chapter 6. The popular semicustom gate arrays are described. Complete custom chips made up from standard-cell arrays, functional-block arrays, and silicon compilers are some of the IC concepts that are presented. These techniques are designed to reduce chip-design turnaround time (and cost) and place increasing emphasis on Computer-Aided Design (CAD) for VLSI. Much of the current thinking about these design possibilities and the description of the requirements for the CAD systems are presented.

The current status of the new developments in high leadcount IC packages is the subject of Chapter 7. A wide variety of both insertion-type packages and surface-mount packages are described.

Chapter 8, "A Look into the Future," discusses the many new techniques that are being considered to develop advanced single-chip VLSI systems. The importance of the new-found freedoms that exist in digital architectures are discussed and the theoretical problem of assessing an optimum architecture is mentioned. The future of silicon circuits, the Silicon Foundry, the place of software in the VLSI Era, the limits of silicon ICs, and a look at recent developments in gallium arsenide ICs are also subjects for this last chapter.

Thomas M. Frederiksen

Acknowledgments

In trying to tell a simplified story of a very complex industry, I often found myself confused and over my head. I am very grateful to one of my best friends over the years, Bob Dort, who, during a crash course, patiently explained many of the details of the evolution in wafer fabrication. Much of this part of the story results from his perception, expertise, and long association with most of the processes and equipment used in silicon wafer fabrication.

The final manuscript also has benefitted greatly from the careful reading and comments of Matthew Buynoski, Milt Wilcox, and Mark Levi.

Intuitive CMOS Electronics

CHAPTER I

Background on Solid-State Electronics

The fabrication of solid-state electronics is rapidly approaching the theoretical limits of silicon processing. It is interesting to quickly review the history of electronics. A review will provide some perspective about the evolution of electronics for those who may be unaware of past efforts and how today's IC processes have evolved over the years.

1.1 THE SIGNIFICANCE OF AN AMPLIFYING DEVICE

Many of the early basic discoveries in electronics would not have resulted in such useful applications if we did not have a way to amplify electrical signals. Even the great theoretical work of James Clerk Maxwell (1831-1879) and the discovery and demonstration of radio transmission by Heinrich Hertz (1857-1894) would have been of limited usefulness.

Great progress in electronics was made possible when researchers built upon the discovery by Thomas A. Edison of the "Edison effect." Edison noticed that, when he placed an isolated metal plate in his evacuated light bulb, he got a shock when he touched a wire lead that was attached to it. This shock resulted because the metal plate was collecting electrons that were being emitted by the filament of the lamp. Edison's discovery was very important for the later development of the vacuum tubes.

J. Ambrose Flemming, around 1904, built on this concept and is given credit for discovering the two-element vacuum tube: the *diode*. The next important step added a *control grid* between the *filament* and the metal *plate*. This structure was called the "audion." This was the discovery of the *triode* vacuum tube; a very important device which could amplify small signals.

Lee DeForest owned the patent on the audion, but there was an unsuccessful legal fight that tried to place credit with Flemming. In 1912, DeForest and his coworkers built a cascade of audions to provide higher voltage gain than was available from a single vacuum tube. The triode and the resulting electronic amplifier were basic to the increased interest and the many practical applications of electronics.

Ways to Provide a Source of Electrons

In this early vacuum-tube-triode amplifier, Figure 1-1, electrons were "boiled off" a heated *cathode* wire within the evacuated glass envelope. (This heated wire was the light-emitting filament of Edison's light bulb. In addition to emitting light, this filament also gave off, or emitted, a steady stream of electrons. This is called *electron emission* or *thermionic emission).* These electrons originated at the surface of the wire of the filament. The thermal energy that was picked up by these electrons from the heated filament raised the energy of the electrons. These energetic electrons could escape from the solid wire of the red-hot filament and jump out into the vacuum that existed within the enclosed glass envelope. (It is sometimes confusing that current flow is *defined* to be in the opposite direction to the electron flow.)

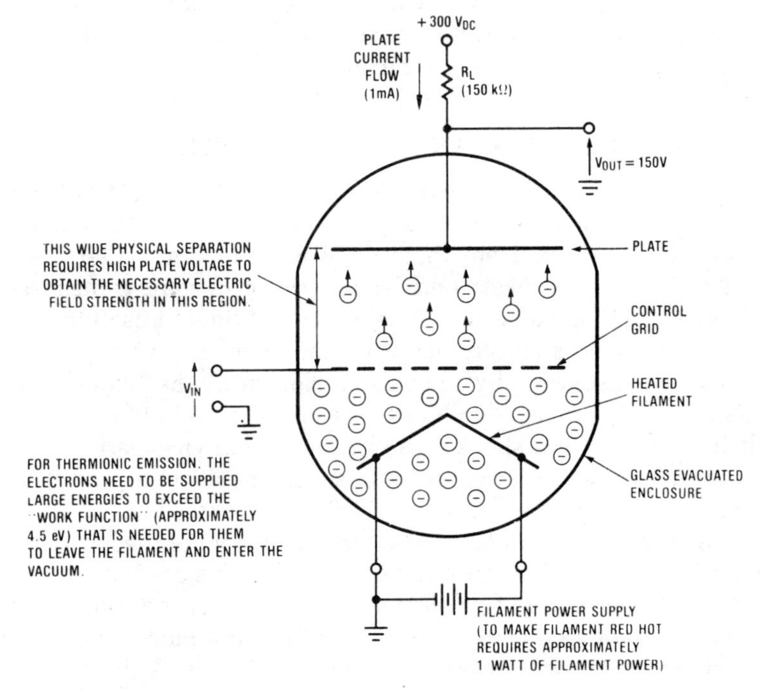

Fig. 1-1 The Vacuum-Tube-Triode Amplifier

The energy needed for an electron to free itself from a solid and jump into a vacuum is called the *work function*. The magnitude of this relatively large energy, approximately 4.5 eV, depends upon the materials that are used for the filament. It also applies to the more efficient materials that are heated by the filament, when a separate cathode structure is used.

It is fortunate that another way of injecting electrons was discovered that does not require these large amounts of energy. This new low-energy PN junction or *solid-state injection* phenomenon, shown in Figure 1-2, is the key to the high *power-efficiency* of solid-state electronics. Heated filaments, with their high power consumption, are no longer needed. Instead, the *free* thermal energy that exists at *ordinary ambient temperatures* is used to *power* this new injection mechanism. However, electrons are not being emitted into a vacuum.

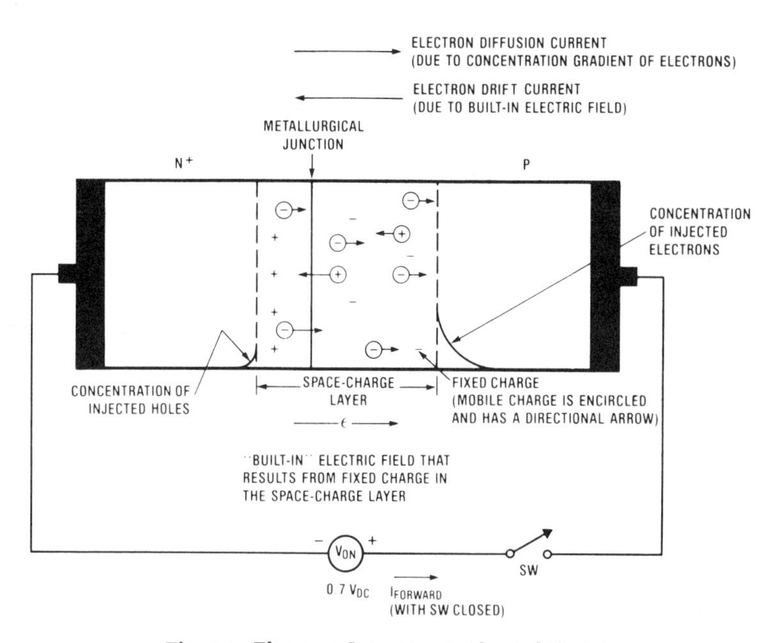

Fig. 1-2 Electron Injection in the Solid State

There is a natural tendency for electrons to diffuse. Electrons diffuse from a point where they are in high concentration, as denoted by the N^+ side of the diode, to a place where they are in a low concentration, the P-side of the diode. Diffusion is the injecting mechanism. Natural diffusion of the electrons is balanced when the switch SW is open and no external voltages are applied. Balance is achieved by an electron *drift current* that results because of the presence of a *built-in electric field*, shown as ϵ in Figure 1-2.

When no external voltages are applied to a PN diode, both of these currents adjust to be *exactly* equal. Therefore, no current is available to flow in an external circuit. When an external forward-biasing voltage is applied to this PN diode, SW now closed, the built-in electric field across the junction is slightly reduced in magnitude. Consequently, a steady diffusion current or electron, for this N^+ P diode, flow results in this conducting diode. If the polarity of the voltage source were reversed, a reverse bias, only a small reverse, or leakage, current would flow in the external circuit.

In any PN diode, each side of the diode is causing a diffusion current flow. At equilibrium, these two diffusion currents, both the electron and the hole diffusion currents, are balanced by two drift current components. These drift current components are also both electron and hole drift currents. Predominantly one type of forward current can be obtained in a forward-biased diode by unbalancing the impurity doping that is used in the fabrication of the diode. The dominant forward current component in an N^+ P diode is electron flow. Conversely, a P^+ N diode would provide dominately hole current flow under forward bias.

This natural source of charge carriers is responsible for the large power efficiencies of solid-state electronics. Lower accelerating voltages can also be used in the *output* circuit because of greatly reduced physical spacings.

Obtaining High Values of Transconductance

A measure of the *quality* of an amplifying device is often expressed as *transconductance* (g_m). This is a measure of how a small change in the input control voltage can cause a relatively large change in the output current of the amplifying device. Transconductance is expressed as

$$g_m = \frac{\Delta I_{out}}{\Delta V_{in}}$$

For an amplifying device, high values of g_m result when the control element is located very close to the source of current carriers, whether electrons or holes. In the case of the vacuum tube, this means that high g_m only results with a very close physical spacing between the control grid and the filament, the source of electrons. The high g_m of the bipolar transistor results because it is the best physical structure of all of the common amplifying devices for keeping the control element, the *base* region, in proximity with the source of current carriers, the *emitter* of the transistor structure.

1.2 A SHORT HISTORY OF AMPLIFYING DEVICES, LOGIC CIRCUITS, AND LINEAR CIRCUITS

Progress in electronic systems depends on such things as size, energy efficiency, and cost of the basic amplifying device. To understand the reasons for the present-day proliferation of electronics, we will consider the technological progress that has been made in improving amplifying devices.

Vacuum Tubes

Vacuum tubes were the first amplifying devices. They were originally very large, relatively high cost, *energy inefficient* glass *bottles* that were approximately 6 inches long and 1½ inches in diameter. These were used in the early radio receivers and were responsible for introducing the *Age of Radio*. (This new communication medium captured the interests of everyone and many people became radio enthusiasts and experimenters.)

Radios appeared in homes and automobiles. But these early radios were large and relatively expensive boxes of electronics that consumed a lot of power. For many years the portable radio was also large and heavy because it needed many batteries of different voltages to allow it to operate. In addition, battery life was short and the high cost of batteries raised the operating expense of portable radios.

Advances in packing density placed multiple individual amplifying devices in a single package—one common glass envelope. The size of the vacuum tube was also scaled down and this resulted in *miniature* tubes that were only 2 inches long and less than 3/4 inch in diameter.

Attempts to *put it all in one package* were also made, even in the old days, that would place all five of the vacuum tubes needed for a standard home radio in one large glass envelope. This development occurred just at the time the germanium transistor was experiencing rapid acceptance. Competition from this new solid state amplifying device stopped most of the efforts to improve the vacuum tube.

Vacuum tubes have suffered from poor control of the initial tolerances of the electrical performance specifications. The electrical specifications depend upon many mechanical tolerances that are associated with the internal structures of a vacuum tube. A special *frame grid* concept was introduced to allow more precise internal element location and also to allow positioning the control grid closer to the cathode surface to raise g_m. This development also came at a relatively late state of maturity of the vacuum tube, and competition from the germanium transistors again tended to reduce interest in improving the vacuum tube.

Germanium Transistors

Solid-state transistor electronics started in 1947 when J. Bardeen and W. H. Brattain, both of the Bell Telephone Laboratories, announced their discovery of *transistor action*. This first transistor came to life when two gold contacts, less than 0.002 inches apart, were made to the surface of a piece of germanium. When one of these contacts was forward biased with respect to the germanium N-type *base* wafer, holes were injected into the wafer. (The name *base* has remained for this transistor lead.) The second contact was spaced close to the first, but was reverse biased with respect to the base wafer. This second contact was, surprisingly enough, found to collect a large portion of the injected holes: this is *transistor action*.

These *point contact transistors* were too fragile for commercial exploitation. However, in 1949, William Schockly, also of Bell Laboratories, introduced the idea for the completely solid structure of the junction transistor. This breakthrough was needed to bring in the *Transistor Era* that started in the 1950s.

The early germanium transistors were made in an unusual, by today's technology, manner. A very small diameter rod of single crystal N-type germanium was sawed into thin wafers. Each of these wafers was the *base* upon which only one germanium alloy transistor was built, as shown in Figure 1-3. The P-type doping for the emitter and collector

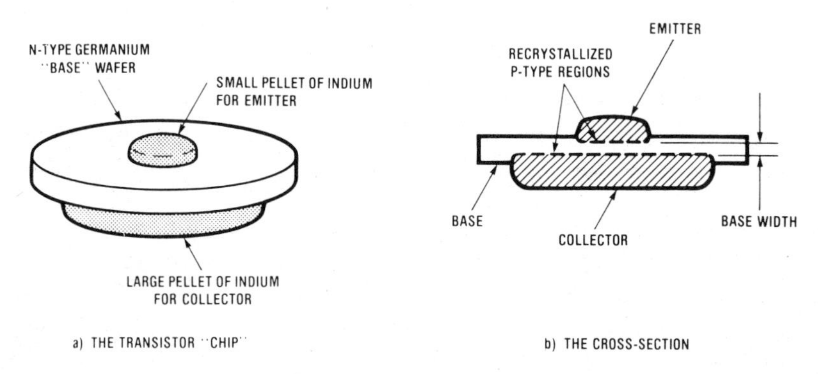

Fig. 1-3 The Germanium Alloy Transistor

regions was achieved by placing small pellets of indium, a P-type dopant, on each side of the germanium wafer. With these indium pellets held in place, the assembly was heated in a furnace. At elevated temperatures, the indium formed an alloy with the germanium, hence the name "alloy transistor." In the recrystallization of this alloy that resulted when the high temperature was removed, the germanium was doped P-type in localized regions on both faces of the wafer. The basewidth for this transistor was the physical separation of these doped regions. This separation was very hard to control in the production of these transistors. There followed a sequence of rapid developments in techniques to provide a controllable and narrow basewidth. Narrow basewidths result in high current gain and high frequency response. Germanium transistors rapidly displaced vacuum tubes as the amplifying device during this emerging age of *Solid-State Electronics*.

Transistors were very small and also very energy efficient. Nothing had to be heated to provide a source of current carriers. Also, high voltage power supplies were no longer needed. Relatively small operating voltages can be used because of the extremely small spacings between the different regions of a transistor. These small spacings result in high values of electric fields with low applied voltages. (Electric field strength is measured as volts per meter of separation.)

The continued reduction in spacing is the reason that the industry will eventually be forced to reduce the value of the present, standard, logic power supply voltage. *Even 5V is too high a voltage* for the small-sized transistors in the upcoming generation of VLSI.

To emphasize the low power requirements of transistors, in the early days, a couple of wires stuck into a lemon were often used in demonstrations to power a simple germanium transistor circuit. The electrolysis taking place in this cell created sufficient electrical power for the transistor circuit.

This low operating power benefit of transistors allowed portable radios to operate from a single, long lasting, small 9V battery. A complete low cost transistor radio would fit into your pocket. Digital computers also no longer required large rooms full of power-wasting vacuum tube electronics. Consequently, the size of computers was drastically reduced, as well as the electrical bill to cool the room.

Silicon Transistors and the Evolution of Logic Circuits

There was a problem with excessive reverse-biased junction leakage at high temperatures. This problem was solved when researchers shifted from small energy-gap germanium to larger energy-gap silicon for transistor fabrication. The higher carrier mobility in germanium with its naturally higher frequency capability was sacrificed in the move to silicon-based semiconductors. (Electron mobility in germanium is approximately three times faster than in silicon and hole mobility in germanium is also a factor of four times faster than in silicon.)

There is current interest in gallium-arsenide (GaAs) as a semiconductor material for high frequency applications in discrete and IC devices. GaAs has a higher energy band gap and higher carrier mobilities than silicon. GaAs is rapidly becoming very useful in new applications and is no longer limited to only Light Emitting Diode (LED) products.

The shift to silicon provided many other natural benefits that resulted from silicon dioxide (SiO_2). This thermally formed, excellent dielectric material is used to encapsulate and to protect the junctions of a silicon transistor. When it was discovered that SiO_2 could also block the diffusion of the impurity dopant atoms that are used to form the N- and

P-type regions within the silicon (a benefit that the oxide of germanium does not possess), the very successful *silicon planar transistor* was born. This occurred in the late 1950s. The key idea of this fabrication technology was to establish the basewidth as the vertical difference between the penetration of the diffusion of the base region and the penetration of a subsequent emitter diffusion (that was located totally inside the previously-diffused base region), as shown in Figure 1-4.

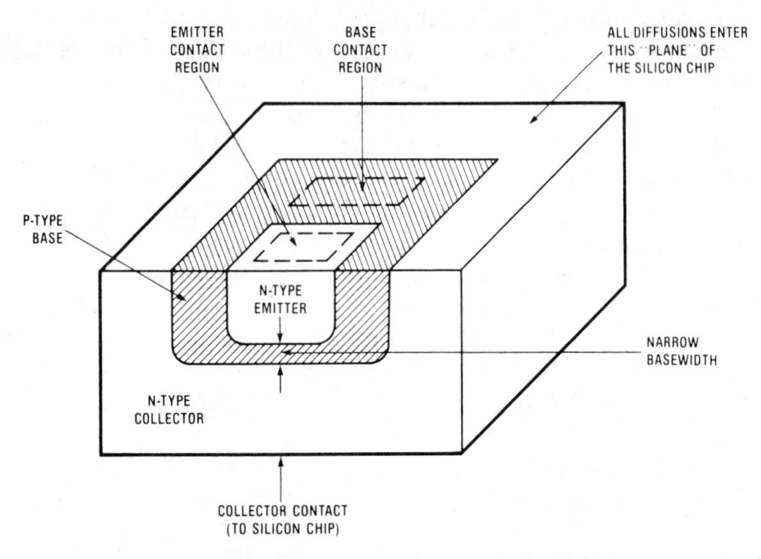

Fig. 1-4 Establishing the Basewidth in a Silicon Planar NPN Transistor

A further fabrication benefit resulted because of the lateral diffusion of both the base and emitter regions. This placed the newly formed PN junctions under a protective SiO_2 surface layer which greatly reduced the leakage current problems that exist when PN junctions are exposed to surface contamination. These easily fabricated, narrow-basewidth, and therefore high-current-gain and high-frequency silicon transistors could operate at high junction temperatures and created competition for the well-established germanium transistors.

This solved the manufacturing problem that was inherent in the silicon alloy transistor because the basewidth was now relatively easily controlled by the differences in the penetration of the base and emitter diffusions. The thickness of the starting wafer and the poorly controlled alloying operation no longer affected the basewidth.

This planar process also allowed many individual transistors to be simultaneously fabricated on one, large diameter, silicon wafer. The costs of processing this wafer were shared by a relatively large number of individual transistors that were simultaneously provided—the "batch fabrication" concept.

The finished wafer was "broken up" by a scribe and break operation or many different sawing techniques. Each transistor "die" was individually packaged, as shown in Figure 1-5.

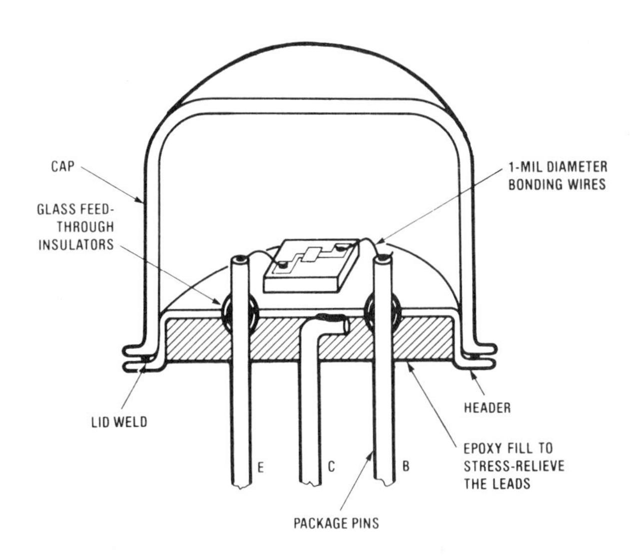

Fig. 1-5 A Transistor Die-Bonded in a Header with the Base and Emitter Wire Bonds in Place

Even in the transistor days there was an economic benefit to the use of large diameter wafers. The limit on wafer diameter has been due to the crystallographic problems of "pulling" large, high quality single crystals (boules) from a high temperature crucible (where the silicon exists in the liquid state). The continued improvements in providing large diameter, defect-free crystals have been a major factor in providing low cost semiconductors.

A benefit of silicon transistors in the design of logic circuits is the increased magnitude of the base-emitter ON voltage that is required to establish a given value of collector current. The forward characteristics of both silicon and germanium diodes are compared in Figure 1-6.

This basic thresholding characteristic of a silicon diode also exists with the base-emitter junction of a silicon transistor. Although a silicon transistor is actually conducting collector current for very low values of base-emitter ON voltage, the resulting collector current magnitude is small enough that it can be neglected in many practical circuits. For example, as can be seen in the logic inverter of Figure 1-7, with a *pull up*

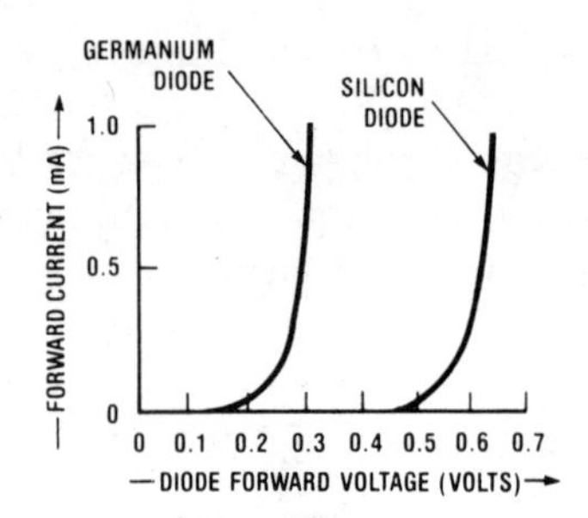

Fig. 1-6 Comparing the Forward Characteristics of Silicon and Germanium Diodes

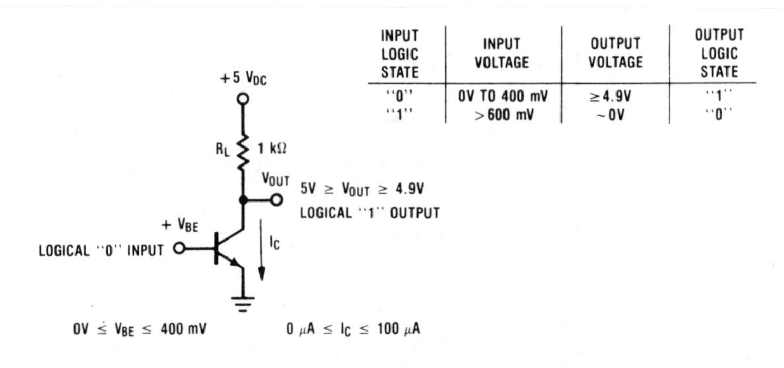

INPUT LOGIC STATE	INPUT VOLTAGE	OUTPUT VOLTAGE	OUTPUT LOGIC STATE
"0"	0V TO 400 mV	≥4.9V	"1"
"1"	>600 mV	~0V	"0"

Fig. 1-7 A Silicon Transistor Logic Inverter That Allows a Relatively High Logical "0" Input DC Voltage Level

resistor as small as 1 kΩ, a collector current flow of at least 100 μA is required to cause the output voltage that exists at the collector of this transistor to be 100 mV less than the 5V dc power supply voltage, or 4.9V. An output voltage as large as 4.9V is still high enough to be "recognized" as a valid logical "1" state. A problem in logic circuitry exists if the logical "1" voltage becomes too small in magnitude to be "recognized" as a valid logical "1" voltage level. Similarly, a logical "0" voltage level should remain close to 0V so that there is no problem of this voltage level turning ON a following logic circuit.

Logic circuits have some "room" for voltage degradation and this is the reason for the *noise margin* of a logic circuit. In contrast to this more forgiving voltage level situation, linear circuits have *no* noise margin. In fact, sometimes the desired signal in a linear system is only 1 μV. This is the reason that much care has to be taken to keep the many natural and man-made sources of electrical noise from contaminating linear circuits.

Silicon transistors have the characteristic of maintaining a low value of collector current at relatively large values of base-emitter ON voltages (400 mV). The silicon transistor has a useful, approximately ON-OFF thresholding or toggling feature that was not available with the smoother turn-ON at low base-emitter voltages that exists with a germanium transistor. This unique thresholding characteristic allowed silicon digital logic circuits to be operated with only a single supply voltage. A single supply was a major simplification because some earlier germanium transistor logic circuits required power supplies of both polarities and used as many as six different voltages.

An ON-OFF thresholding characteristic that is useful for digital circuits also exists with the enhancement-mode MOS transistor. This was not available with the naturally depletion-mode operation of the JFET transistors. We will have more to say about this in the next chapter.

An early logic circuit design was based on the fact that the *collector voltage* of a silicon transistor (that was driven into saturation), was *smaller* in magnitude *than the base-emitter voltage that was necessary to turn ON a following transistor* (that had its base tied to this collector), as shown in Figure 1-8. This relatively simple logic circuit was called *Direct Coupled Transistor Logic* (DCTL). Although this circuit could operate properly when only one transistor was driven by any one collector, it suffered because of unequal current sharing when the bases of additional transistors were tied to the same collector. (The ability of one logic circuit to provide a logic output state that can drive many following logic circuits that require this logic signal as an input is called *fan out*.)

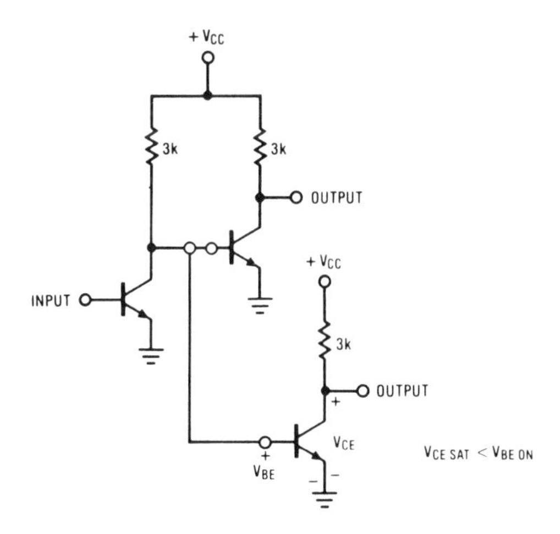

Fig. 1-8 Direct Coupled Transistor Logic (DCTL)

It is impossible to fabricate many individual transistors that have identical thresholding characteristics. This was the major problem with DCTL. There was therefore no way to guarantee that the many transistors at the inputs of the logic circuits that could be connected in parallel would all perform in an identical manner. That is, these input transistors would not equally share the drive current that was available from the pull-up resistor of the logic gate that was driving all of these inputs when the logic at the input was in the "1" state.

This "current hogging" problem of DCTL was solved by adding *ballasting* resistors at the inputs of the logic circuits. This innovation, Figure 1-9, was called *Resistor Transistor Logic* (RTL). RTL was one of the early IC product offerings and appeared in the early 1960s. Although these input resistors did serve to equalize the available drive current, they unfortunately slowed down the circuit response time. This loss in switching speed is never desired in a logic family but resulted because an input voltage change was somewhat isolated by these ballast resistors. The voltage directly at the inputs of the inverting transistors could not be changed as rapidly when the input ballast resistors were added.

The resistors needed for the RTL circuits are part of the IC technology. Resistors are realized by placing a stripe of the P-type base diffusion into a separate epi tub. Contacts to the ends of this stripe access the resistor. This separate epi tub is tied to the most positive supply voltage, +5V for a logic circuit, to keep the P-type resistor reverse-biased with respect to the epi tub in which it is located.

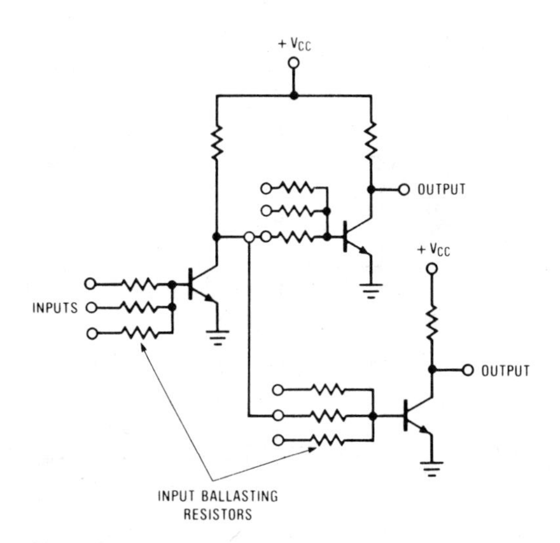

Fig. 1-9 Resistor Transistor Logic (RTL)

The move from the production of single transistors to complete multi-transistor circuits, the early digital logic Integrated Circuits (ICs), required a way to electrically isolate the individual transistors. It was intended that these new ICs would take advantage of the performance and processing benefits of the successful silicon-planar transistors.

In fabricating a single transistor, a heavily N^+ doped wafer is used as the starting material. A thin layer of much lighter, N-type doped silicon is grown on the surface of this substrate by vapor-phase *epitaxial deposition*. The resulting *epi layer* growth propagates the underlying crystal structure of the substrate and provides the desired high resistivity collector region for the NPN transistor, as shown in Figure 1-10a.

In fabricating the transistors that are required for an IC, only the collector regions need to be electrically isolated. As can be seen in Figure 1-10a, the base region, which contains the emitter, is diffused into—and is isolated by—the collector region. The electrical isolation of the N-epi

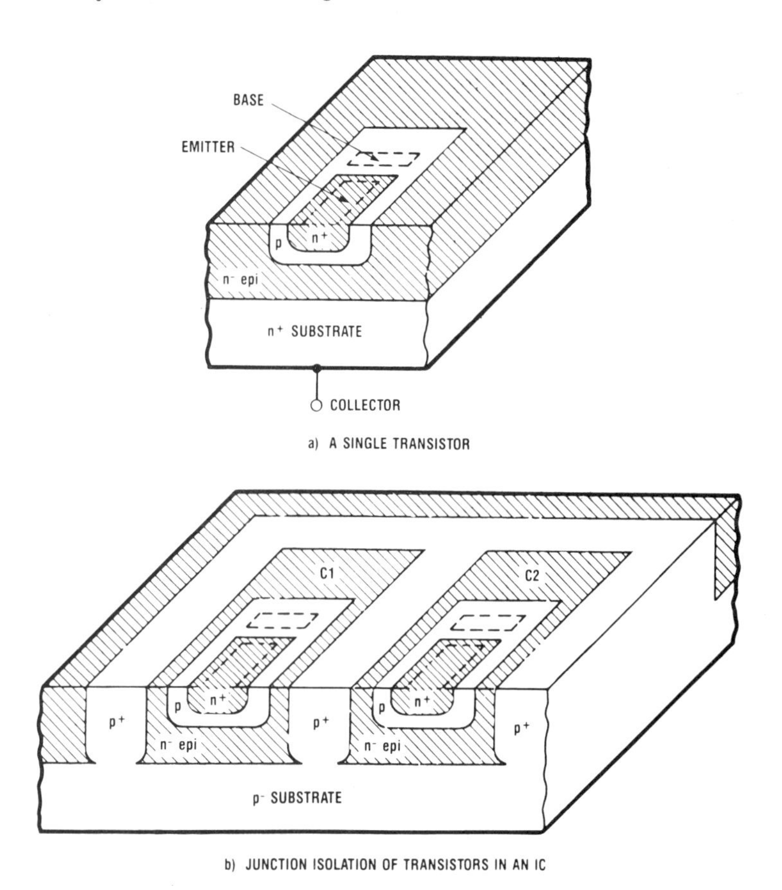

Fig. 1-10 The Multi-Transistor Junction Isolation Used in ICs

transistor collectors in an IC is obtained by forming PN junctions between them that will always be reverse-biased.

To achieve this reverse-biased *junction isolation* requires that the N-epi layer of an IC be grown on the surface of a P-type substrate, as shown in Figure 1-10b. This substrate will be electrically connected to ground, or the most negative power supply. The individual collectors of the NPN transistors will all operate at more positive voltages. This will provide the desired reverse-biased isolating PN junctions across the bottom surfaces of the individual collector regions.

To electrically isolate the edges of these individual transistor *epi tubs*, a special P-type *isolation diffusion* is used to break up the surface N-epi layer into small, transistor-sized regions, as shown in Figure 1-10b. This rich P-type isolation diffusion is driven completely through the N-epi layer and places reverse-biased PN junctions along all four edges of each of the collector regions.

The next logic circuit improvement used diodes to solve the problem of connecting many logic circuits together and was called *Diode Transistor Logic* (DTL). This type of logic is shown in Figure 1-11.

DTL solved the current hogging problems of DCTL and the speed problems of RTL. Consequently, DTL became a very popular logic family. The heavy use of diodes in these DTL circuits stimulated the semiconduc-

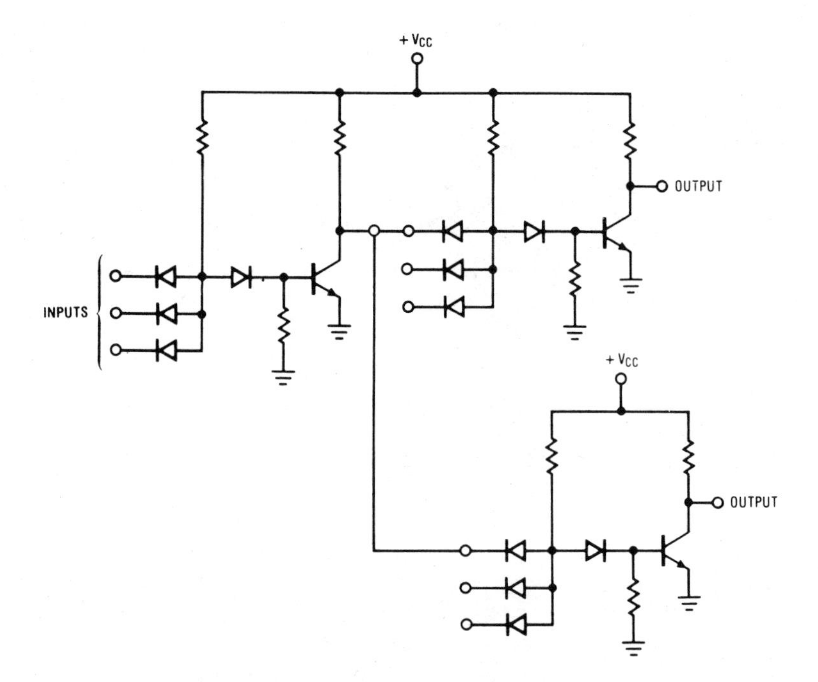

Fig. 1-11 Diode Transistor Logic (DTL)

tor industry. Very large quantities of discrete diodes were supplied to satisfy this newly created demand. (The high cost of the early IC digital logic circuits—$50 to $200 each—forced many designers to continue using discrete DTL logic circuits.)

As integrated circuits developed from their first introduction in 1959, it was a relatively short time before a way was found to supply the diodes that were needed in DTL by making use of an additional transistor. This discovery introduced a strange-looking, new, multi-emitter transistor symbol to the industry, as shown in Figure 1-12.

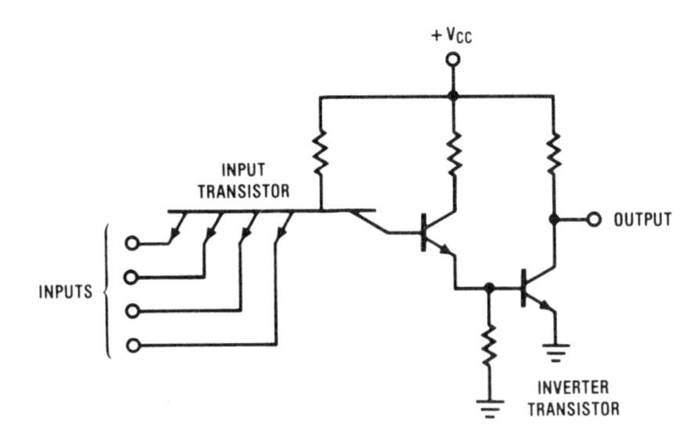

Fig. 1-12 Transistor Transistor Logic (TTL)

This was not only a die-area efficient way to provide the input diodes that were needed in DTL, but it also offered a performance benefit: there now was faster switching because of the *transistor action* that exists within the input transistor that replaced the input diode network. For example, when any one of the inputs drops to the low voltage state, the multiple emitter input transistor becomes active. This is because base current is supplied by the resistor connected from the base of this transistor to the power supply line. The collector of this input transistor can now rapidly remove charge from the base of the following transistor. This more rapid base charge removal reduces the time delay in the transition to the OFF state of the output transistor and provides a circuit benefit when compared with the performance DTL. This new logic was called *Transistor Transistor Logic* (TTL or T^2L). It has been the workhorse of IC bipolar logic since the early 1960s and is still in production today.

A problem inherent in T^2L is that the inverter transistor is allowed to saturate. To reduce the recovery time from saturation, T^2L circuits use *gold doping* in the wafer fabrication. Gold is diffused into the silicon wafers to provide *recombination sites*. This aids the removal of the excess carriers that exist because the transistor was driven into saturation.

These excess free carriers must be removed before the OFF state of the transistor can be achieved.

A final step in the evolution of T^2L was to provide clamping diodes across the base-collector junction of the Inverting Transistor, Figure 1-13,

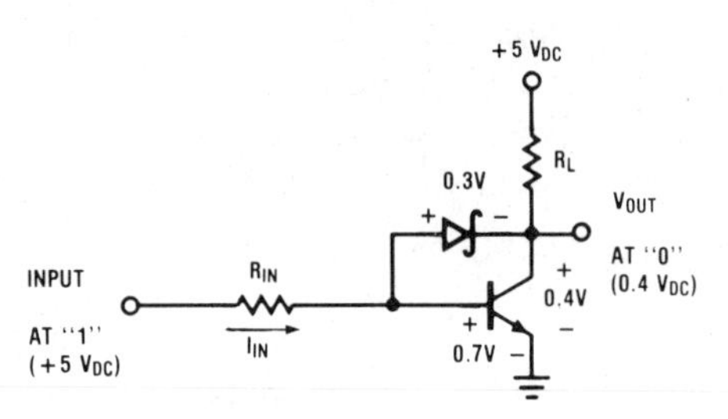

Fig. 1-13 Using Schottky Clamp Diodes to Prevent Saturation

that would divert the excess base drive current into the collector and thereby prevent saturation. This circuit modification is called *Schottky clamped T^2L* because Schottky diodes are used, and has been a very popular bipolar logic family. Gold doping is not used for this logic family, but a rather complex metal system is required to form the Schottky diodes. These diodes are created when a special metal is in contact with the silicon. They have the benefit of a very fast switching time. The inherent speed advantage of this nonsaturating logic circuit has allowed the power consumption to be reduced in a family of Low-power Schottky (LS) logic circuits that still provide standard T^2L speeds.

The early bipolar IC logic circuits consisted of the circuit complexity that existed on popular 4×6 inch printed circuit cards that contained discrete transistorized logic circuits. This initiated the *Small Scale Integrated (SSI) Circuit* (3 to 30 gates per chip) *Era* in the early 1960s. These relatively simple ICs became very useful building blocks for digital systems.

As processing yields increased, the semiconductor industry responded to the ever-constant demand for increased circuit complexity on a single chip. Thus, the *Medium Scale Integrated* (MSI) Circuit (30 to 300 gates per chip) trend started in the mid to late 1960s and more complete functions such as counters, shift registers, and other larger groupings of logic circuitry were made available to the system designers.

The demands for still more complex logic strained the capabilities of the relatively complex bipolar wafer fabrication processes and the reduced yields of the larger logic chips (which are more vulnerable to defects) forced the technologists to find a simpler way to realize IC digital circuits. This motivation for increased logic circuit complexity stimulated the interest in a less complex transistor: the *Metal Oxide Semiconductor* or *MOS Field-Effect Transistor* (MOSFET).

The Linear IC Problems

The interest in linear IC circuits in the early 1960s followed on the heels of the early digital ICs. As a result, the early linear IC designers inherited the low breakdown bipolar processing that was used to fabricate the 5V logic circuits. The linear design problem was therefore: "How can linear circuits can be produced when one only has NPN transistors and relatively low valued and poor initial tolerance diffused resistors?"

The new circuit design concept that evolved was that linear circuit designs could now make use of a large number of transistors—but they were all only one type: NPNs. Large-valued coupling or bypass capacitors did not exist; and also, inductors were not available (these facts haven't changed).

A major step forward came from the early realization that a very significant circuit performance advantage resulted from the *natural component matching* of both the resistors and the transistors that were simultaneously fabricated and were also in proximity on an integrated circuit. Although the initial tolerance was poor, the components would have good ratio matching. The goal was to design circuits with only ratio dependence. This component matching was good enough to allow a transistor to be voltage-biased by the base-emitter forward voltage drop (the ON voltage) of a diode-connected reference transistor, as shown in Figure 1-14. (This early discovery is believed to be first recognized and used in an IC design by James E. Solomon in the late 1950s.)

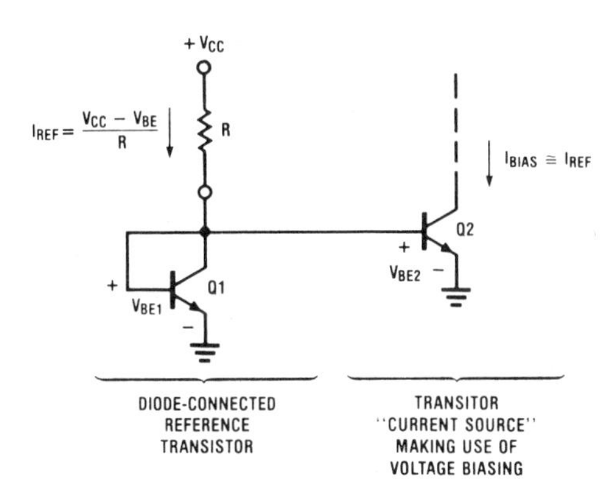

Fig. 1-14 Using Matched IC Transistors in IC Circuit Design

The collector-base short used on the reference transistor, Q1, is a way to force Q1 to carry I_{REF}. The V_{BE1} that is needed for Q1 to support a collector current of I_{REF} naturally results and is then directly applied as a voltage bias for Q2. The matching of these IC transistors forces Q2 to also support a collector current of I_{REF} (neglecting the base currents and the

Early effect: the effects of the different collector voltages). A useful property of this basic biasing circuit is that the relative emitter areas of Q1 and Q2 can also be scaled to produce unequal current replication in Q2. For example, if Q2 were to have a larger emitter area, it would then conduct a proportionately larger collector current. (This results because the forward voltage drop of the base-emitter junction of a transistor, that is needed to support a given value of collector current, directly depends on the emitter area.)

Another useful linear IC circuit trick, which also makes use of the matching that can be achieved between transistor base-emitter forward voltages, makes use of ΔV_{BE} circuits. These have been very useful in providing voltage references, in temperature compensation schemes, and in output current-limiting circuits.

The early IC voltage regulators usually made use of the 7V BV_{EBO} of the NPN transistors or the less reproducible 5V (N^+ emitter-to-P^+ isolation junction) zener breakdown voltage that also exists within the standard linear IC process. Both of these zeners require relatively large input voltages, drift with time, are noisy, and have positive temperature coefficients (+TCs). The natural negative temperature coefficient (−TC) of the base-emitter forward voltage of a transistor was used in combination with these zener voltages to provide many clever circuit designs that would provide an overall voltage reference with essentially a zero temperature coefficient (0TC).

A need existed for a lower-magnitude reference voltage that did not require a high input voltage just to break down a zener diode and also had less noise and better long-time stability. It had been known that the −TC of the forward voltage of a transistor (usually assumed to equal −2 mV/°C) did actually depend on the magnitude of current that was flowing in the transistor (or more precisely, the current density: the relative area of a transistor is as significant as the relative magnitude of the emitter current). For example, a transistor carrying a larger value of current would have a smaller value of temperature change in V_{BE}.

The thing that was new with ICs was that two different, but matched, transistors could be placed within a circuit such that their V_{BE} voltages would subtract. Then, if these transistors were operated at different current densities (obtained by ratioing the emitter areas and/or the current flow), a ΔV_{BE} voltage would result that had a +TC (because the base-emitter voltages of these two transistors will change with temperature at different rates). This was again offset with the −TC of a V_{BE} to obtain an overall low-voltage reference, with essentially a 0TC.

These have been called band-gap voltage references because the early circuits would sum 0.6V of ΔV_{BE} (which has the same percentage change per degree centigrade as V_{BE} but is of the opposite sign) with 0.6V of V_{BE} to obtain a 0TC reference voltage of 1.2V. This, in electron-volts, is the

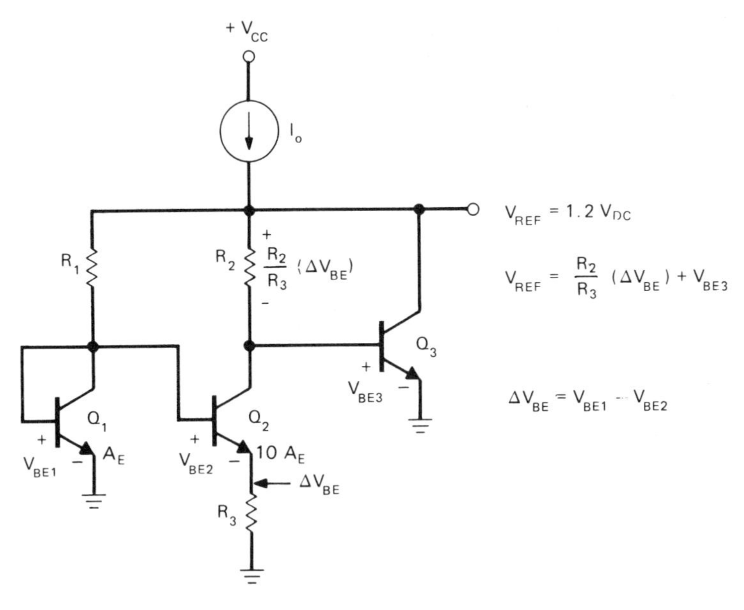

Fig. 1-15 Basic Band-Gap Voltage Reference

band-gap energy of silicon. This removed the requirement for a high input voltage, reduced the noise, and improved the long-time stability of these band-gap reference voltages, as compared to the earlier zener-diode derived reference voltages.

One of the early 0TC voltage reference circuits that made use of this concept is shown in Figure 1-15. The two transistors that provide the ΔV_{BE} voltage are Q1 and Q2. Notice that Q2 is a much larger geometry transistor. Q2 has 10 times the emitter area of Q1 or consists of 10 transistors (the same size as Q1) all connected in parallel. In addition, the emitter current of Q1 is designed to be larger than that of Q2. Both of these factors cause the emitter current density of Q1 to be larger than that of Q2, so V_{BE1} is larger than V_{BE2}. The difference between these voltages, the ΔV_{BE} voltage, appears at the emitter of Q2 (as shown on the figure). This voltage is then amplified by the ratio of R2/R3 and appears across the resistor R2. The reference voltage produced by this circuit, V_{REF}, is seen to be the voltage that is developed across R2 added to the base-emitter voltage of Q3, V_{BE3}, as shown on the figure. The amplified ΔV_{BE} term has a +TC and the V_{BE3} term has a −TC. When both of these terms are designed to be equal (each will be about 600 mV) V_{REF} will have essentially an 0TC.

MOS versions of this band-gap voltage reference technique are used which use the difference in the threshold voltages of two MOS transistors. A ΔV_{BE} voltage has also been used as the reference to provide for the biasing currents in the all-CMOS op amps that will be described in Chapter 5.

Linear circuit designs always have taken advantage of the component matching that is inherent in IC fabrication. New circuit design tricks reduced the dependency on capacitors and inductors, but the lack of a complementary transistor (a PNP) created a large circuit design problem. Therefore, the discovery of a lateral PNP transistor, and later a vertical PNP, greatly aided the linear circuit designers.

The linear IC designers have always been plagued because the standard IC process does not provide a "good" PNP transistor. Many ingenious substitutes have been used and this problem remains today (although more complex processes, that will be discussed later, are now solving this problem). It is interesting that the bipolar digital ICs did not require a PNP and it is only the more recent shift to a CMOS process that has provided the benefits of a good PNP (the P-channel MOSFET) for the design of low current-drain logic circuits.

The shift by the digital designers to CMOS processes has caused a corresponding interest in CMOS for linear ICs. A good PNP is inherent in the advanced CMOS processes. The existence of complementary transistors, where both are high-frequency devices, provides a major simplification to the circuit designer, whether linear or digital. While CMOS solves the problem of obtaining a good PNP, it presents a few new linear circuit design problems of its own. We will consider some of the solutions to these problems later in Chapter 5.

The performance of linear circuits depends on the care and the "cleverness" used in the chip layout design and also the care and the "cleverness" used in wafer fabrication. (The performance of digital circuits is more immune to both of these factors.) A successful linear IC product therefore depends on the efforts of many people: the circuit designer, the mask designer, the wafer processing people, and the product engineer. Many times special wafer processing modifications are made to increase the yield of a particularly sensitive linear IC.

The ion implanter (a replacement for a diffusion furnace—that we will consider in Chapter 3) has provided not only well-controlled, high-valued resistors, but also a process-compatible, P-channel JFET transistor. To support special linear designs, ion implanters have also been used to fabricate compatible PNPs and high-speed NPNs. The discovery of these additional circuit elements has greatly aided the ability to provide high performance linear ICs.

The Evolution of Silicon Wafers

To reduce the handling costs of each IC chip, the original ¾-inch (19 mm) diameter wafers (also called "slices") that were used for the early ICs were quickly increased to 1½ inches (38 mm) in diameter. Technologists have been continuously working to provide large diameter wafers. In 1971, 3-inch (76 mm) wafers were introduced to the IC fabrication lines. Increasing the diameter from ¾ inch to 3 inches increased the wafer area by a factor of 16:1. The 3-inch wafers were in high volume production until the first of 1980 when 4-inch (100 mm) wafers started to take over. Then, in late 1982, 5-inch (125 mm) wafers went into high volume IC fabrication lines. This increase in wafer size is shown in Figure 1-16, where wafers from 1½ inches to 5 inches are shown.

As the wafer diameter increases, the thickness of the wafer also must be increased to prevent breakage from handling during wafer processing.

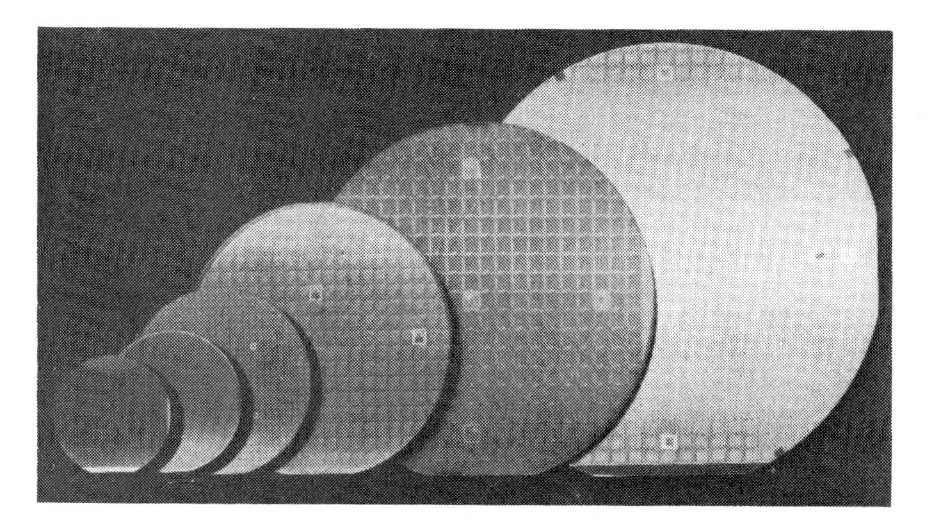

Fig. 1-16 The Increasing Sizes of Silicon Wafers

For example, the 8 mil (0.008 inches) thickness of the 1½-inch wafers was increased to 25 mils for the 5-inch wafers. These thick wafers are thinned down to approximately 14 mils at the completion of wafer processing, prior to packaging. This is a difficult operation and is done by mechanically abrading the back surface of the wafer. This must be done without degrading the sensitive completed circuits that exist on the opposite side of the wafer. (A mistake at this point is very costly.)

By the end of 1983, 6-inch (150 mm) wafers were in full production and suppliers are now working on 8-inch (200 mm) wafers. These 8-inch wafers are so large that to relieve the film stresses that are created by the layers that are used in IC fabrication requires a hole to be placed in the center of the wafer. It is not clear that these should still be called "wafers" because now they resemble giant "flat washers." This is an amazing accomplishment by the crystallographers. The huge crystals that must be grown to supply these wafers require an enormous quantity of silicon and must be produced without crystallographic defects. The amazing thing is that technologists are projecting the future need for 10- (250 mm) and 12-inch (300 mm) diameter silicon wafers! The jokes about silicon wafers becoming as large as a pizza may actually come true.

Each production line change to handle larger diameter wafers usually obsoleted essentially all of the existing fabrication equipment. The costs for the larger wafer-capability machines also increased; but, in addition to handling the larger diameter wafers, the performance of each machine must be increased to maintain high yields. For example, the move to 6-inch diameter wafers also brought robotics to the IC wafer fab lines and a high degree of automation. These changes are responsible for the large increase in the cost of installing a new wafer fab module.

The consumption of silicon wafers in the U.S. exceeds 70 million wafers per year. To appreciate what 70 million wafers amounts to, if we assume that the average wafer is 15 mils (0.015 inches) thick, this would represent a stack of wafers over 15 miles high! The total weight of our annual requirements for silicon (assuming an average wafer diameter of 4 inches) would be slightly over 500 tons of *ULTRA-PURE* silicon that has been derived from an extensive sequence of chemical refinements. The semiconductor companies do not supply this total need. Some buy all of their wafers from outside vendors. Thus, the production of silicon wafers has become a major industry.

Annual wafer consumption is remaining relatively constant because the size of each wafer is increasing at the rate of approximately 1 inch every 5 years. The use of larger diameter wafers reduces the labor costs per individual circuit that is produced and also reduces the amount of silicon that is wasted around the perimeter.

The area of a wafer is proportional to the square of its diameter. Therefore, there was a larger benefit in going to 3-inch wafers from 1½-inch wafers (9 to 2.25 or a 4:1 increase in area) than there now is in going from 5-inch to 6-inch wafers (25 to 36 or a 1.44:1 increase in area).

Between 1953 and 1975, technological improvements in silicon crystal growing resulted in significant wafer cost reductions. These costs reached a minimum in 1978. The increasing energy costs have now reversed this trend and the cost of silicon wafers is steadily increasing.

It is interesting that a controlled amount of oxygen is purposely left in the silicon in the manufacture of silicon wafers. This provides *point defects* in the bulk that enhances the gettering of contaminants. (We will have more to say about getters in Chapter 2.)

Bipolar Enhancements

Although there has been considerable effort expended to introduce and improve MOS processes, the bipolar processes have also undergone a steady sequence of improvements. We have collected a few of the more interesting bipolar developments in this section and will now take a look at them.

FIELD THRESHOLD ADJUSTMENT. The relatively light doping, the high resistivity, of the N-epi layer of bipolar linear IC products creates problems with parasitic MOS transistors that can degrade the circuit performance. This problem is made worse in plastic packages where there is usually a higher level of silicon-contaminating ions than exist in hermetic packages.

A process modification has been used to solve this contamination problem. It is called *Field Threshold Adjustment*. The basic idea is to add a final N-implant step near the end of the process. This implant is used to carefully increase the surface doping concentration of the N-epi without significantly reducing the maximum operating voltage of the finished product. This richer surface doping is more difficult to invert, under parasitic P-channel MOS action and, therefore, linear circuits that incorporate Field Threshold Adjustment are less affected by stress testing in plastic packages. This process enhancement has provided a large measure of improved reliability for plastic encapsulated linear products and is therefore a significant step toward producing a "bullet-proof" die.

ADVANCED BIPOLAR LOGIC FAMILIES. Many customers are unaware of the fierce competition that exists among the product groups within the same IC manufacturing company. For example, the large speed improvements that were made in the NMOS products by the use of device scaling (going to smaller feature sizes), shallow junction depths, completely ion-implanted processes, and the use of oxide isolation (we will look at these improvements in Chapter 2) were quickly adopted by the competing bipolar product groups. This resulted in smaller geometry bipolar transistors that had less stray capacitances and narrower basewidths. Therefore the f_τ (the frequency performance) of the bipolar transistors was raised from approximately 800 MHz to greater than 5 GHz. These bipolar *Oxide Isolated Silicon* (OXISS) processes also make use of two layers of interconnect metal to increase the component density (much like a double-sided printed circuit board).

The circuit performance improvements that result from the use of these new bipolar processes has breathed new life into the bipolar logic product groups. These OXISS processes have been responsible for the introduction of additional high-performance bipolar logic families: the 74AS *(Advanced Schottky)* and the 74ALS *(Advanced Low-Power Schottky)* series. For example, the 74ALS logic family achieves propagation delays that are approximately one-half that of 74LS and yet the new 74ALS products draw one-half the power of 74LS. In addition the Fairchild Advanced Shottky Technology (FAST) has provided strong competition with logic performance that lies between AS and ALS. (These performance combinations place these logic families in competition with the advanced CMOS logic family, 74HC, which we will consider in Chapter 5.)

A performance comparison of these advanced bipolar logic families is shown in Figure 1-17. It will be the customer orders, rather than the IC manufacturers, that will determine the acceptance and viability of these new IC logic products. In fact, there are specific system application examples where each logic family has advantages.

FAMILY	PROPAGATION DELAY (ns)	Gate POWER (mW)
LS Std	8	2
ALS Std	4	1.3
ALS Buffer	4	3
54S	3	20
FAST	2	4
AS	1.5	7.6
AS Buffer	2.0	8

$V_{CC} = 5V, C_L = 15\text{ pF}$

Fig. 1-17 Performance of Advanced Bipolar Logic Families

INCREASING THE IC POWER LIMITS. Bipolar power ICs have problems competing with the performance that can be realized with discrete bipolar power transistors. The discrete power transistor, shown in Figure 1-18, uses only a relatively thin, high-resistivity epi layer for the collector region. This epi is deposited on a very heavily, N$^+$-doped, substrate that

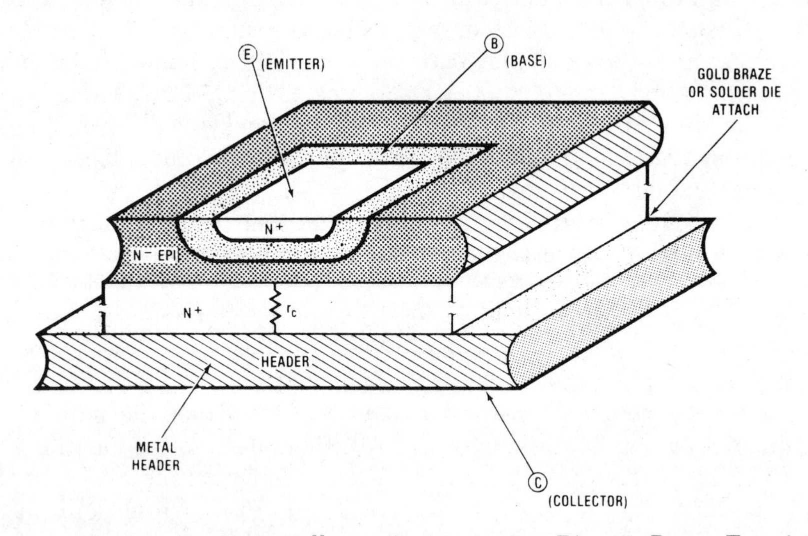

Fig. 1-18 The Low Extrinsic Collector Resistance in a Discrete Power Transistor

comprises most of the thickness of the transistor chip. The vertical transistor action allows the collector current to flow in this low-resistance N^+ substrate, shown as r_C on the figure. This reduces the extrinsic resistance in the collection region (r_C) and, therefore, allows large values of collector current flow with only a small collector-to-emitter voltage drop.

The requirement for isolated N-epi *tubs* in an IC layout greatly raises the extrinsic resistance in the collector region, as shown in Figure 1-19. No longer can the collector contact be made to the bottom surface of the chip as in discrete transistors. Instead, a *top-side collector contact* must be used for each transistor.

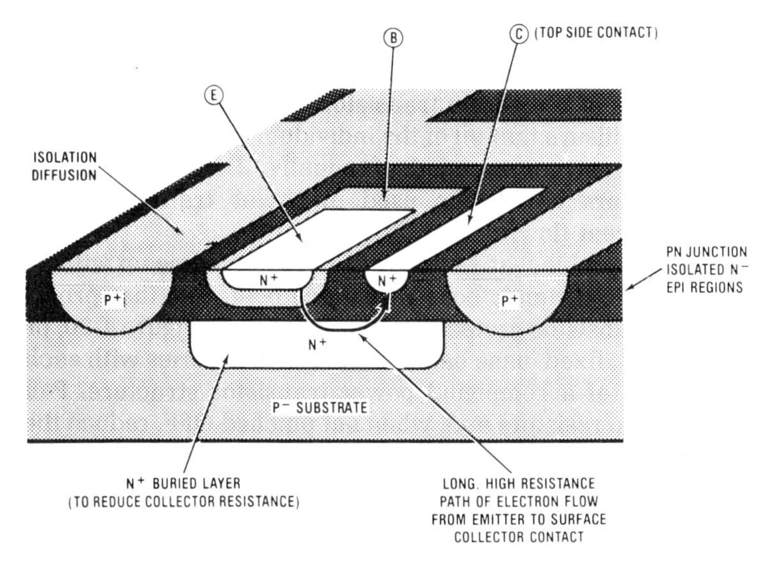

Fig. 1-19 The High Resistance Surface Collector Contact in an IC

The injected emitter current, the electrons, initially flow vertically downward as in a discrete power transistor. This current then must flow horizontally through a special, relatively low resistance, N^+ *buried layer* before it must turn around and then flow vertically up to an N^+ contact diffusion on the top surface of the chip. This long, high-resistance path greatly adds to the extrinsic collector resistance and this increases the collector-to-emitter voltage that results at high collector current flow.

A special process modification has been used, Figure 1-20, that combines a *power transistor process* with a linear, bipolar, IC process to extend the high power limits of ICs. The cross-section shown in this figure indicates how two epi layers are used to simultaneously provide the isolation requirements of the IC portions of the chip along with the low r_C requirements of the power transistor. This new process has allowed 10A ICs to be introduced.

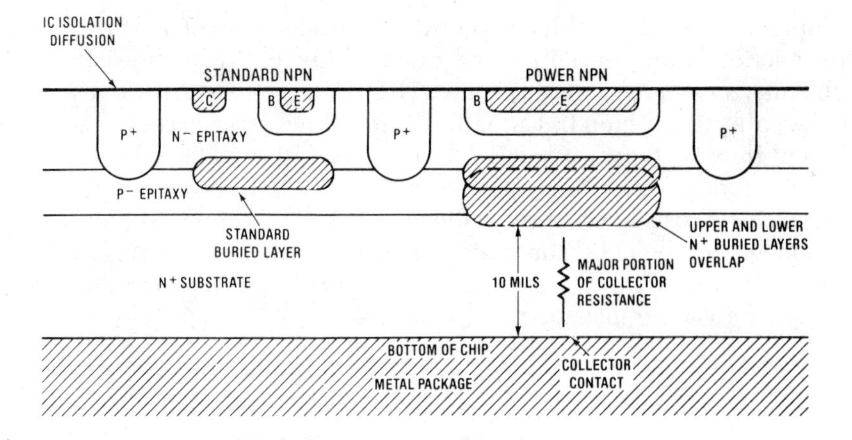

Fig. 1-20 Combining a Power Transistor with an IC

IMPROVING A POWER TRANSISTOR. It is interesting that IC techniques have also been combined with a power transistor to extend the high power limits of a discrete transistor. In power transistor design, a way is needed to insure that all of the individual small sections of the composite transistor structure conduct equally. Both base and emitter ballasting resistors have been used in power transistor designs to equalize the current flow.

Bob Widlar, an innovative linear IC designer, came up with a way to provide automatic *dynamic base resistance ballasting* that provides improved power transistor protection. This technique, Figure 1-21, places a maximum-valued, fixed base ballasting resistor in series with each of the individual bases of a compound power transistor structure. P-chanel JFETs shunt these resistors and, when not pinched-OFF, reduce the value of the base ballasting resistance. Notice that the gates of these JFETs tie to the collector of the NPN power transistor. Higher values of collector voltage tend to turn these JFETs OFF, and, therefore, to raise the value of the base ballasting resistance. This is the desired action, because more ballasting resistance is needed to insure equal current flow at high values of collector

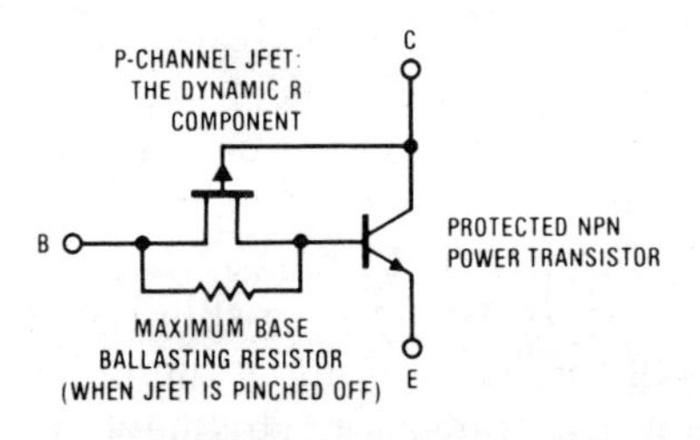

Fig. 1-21 Dynamic R-Ballasting in a Power Transistor

voltage. In addition, the value of both of these ballasting resistors, the fixed and the JFET or *dynamic resistor*, have a relatively strong positive temperature coefficient of resistance. This not only increases the protection at high power levels, it also allows this ballast protection to respond rapidly enough to prevent the formation of localized *hot spots* that are responsible for initiating a "secondary breakdown" mechanism.

The most interesting aspect of this dynamic R-ballasting concept is the simplicity of the chip layout, shown in Figure 1-22. The large white square in the center of this figure is one of the many common base regions. Each base region is connected to the vertical base stripes by narrowed horizontal sections that provide the diffused, fixed base resistors. (These resistors establish the maximum base ballast resistance value.)

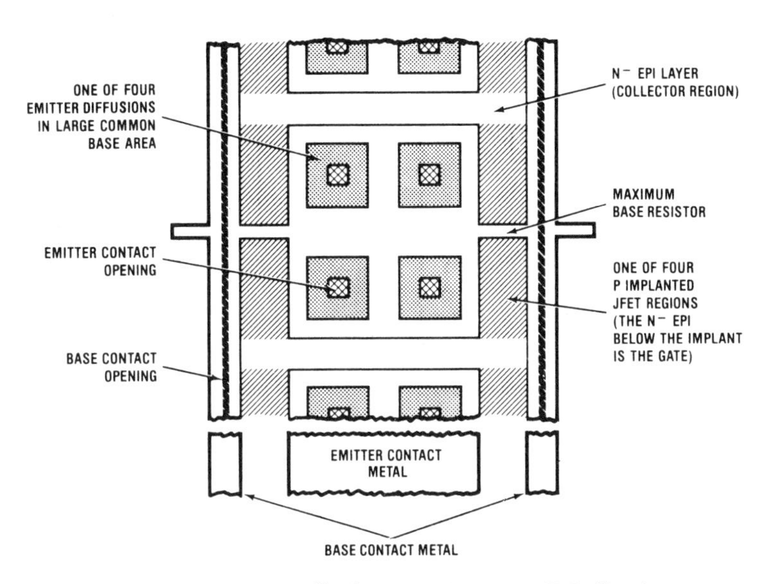

Fig. 1-22 A Basic Cell Showing Dynamic R-Ballasting

The P-channel JFET is easily achieved by using a P-implant that bridges from the large base area to the vertical base stripes. The gate for this JFET is the underlying N-epi collector region. This epi layer was not completely inverted by the shallow P-type surface implant and therefore still exists under the P-implanted channel region. This N-epi region is also the collector of this NPN power transistor.

A wide stripe of aluminum will contact all of the individual emitters of this transistor structure. Thinner base metal stripes go along each side. The collector contact for this NPN power transistor is the substrate of the die and the electrical contact is therefore provided by the die attached to the header, as in a conventional silicon power transistor.

Dynamically, R-ballasted power transistors are available that have an $f_T > 40$ MHz, exhibit no forward-biased secondary breakdown, safe area, at greater than 200W and 200V, and have a V_{SAT} of less than 1.5V at 12A.

DYNAMIC SAFE-AREA PROTECTION. Bob Widlar's more recent contribution to power ICs is a novel technique that is used to increase the power capabilities of a bipolar power transistor by an order of magnitude. The advantage is that the power dissipation rating of a transistor can be custom tailored, and the junction temperature can be more rapidly and more accurately determined, so that the fullest potential of a transistor can be achieved. This is in contrast to the usual technique of making use of heavy deratings that are based on the worst-case transistor that is possible in production (and packaging) and the use of temperature sensors that do not accurately detect the high-temperature regions of a power transistor. Dynamic safe-area protection is especially beneficial when operating at high voltages and under pulse conditions.

This new power transistor protection technique has been used to design an IC power op amp that can provide plus and minus 10 amps to a load with voltage swings of plus and minus 35 volts. This power op amp can provide 150 watts of sine wave power into a 4-ohm loudspeaker, a power level not previously associated with an IC.

In the past, thermal limiting and maximum-junction-temperature shutdown circuits have been used to protect power IC products such as power amplifiers and voltage regulators. Unfortunately, the power capabilities of these IC products have had to be derated because there has been poor thermal coupling between the location of the junction-temperature sensing device and the internal regions of the power transistor where the high junction temperatures exist. To compensate for this inadequate thermal coupling, special circuitry has been used that reduces the magnitude of the maximum current that is allowed to flow as the collector-to-emitter voltage increases. This has been called "foldback" current limiting.

The safe-area curve of a bipolar power transistor places the ultimate limit on the combinations of collector-emitter voltage and collector current that can be simultaneously allowed without triggering the well-known, destructive, safe-area (or second breakdown) transistor failure mechanism. Typically, large values of current are allowed only in combination with low values of collector-emitter voltage. Again, special circuitry has been used that senses the collector-to-emitter voltage that exits across the power transistor and reduces the magnitude of the current that is allowed as the collector-emitter voltage increases. This has been called safe-area protection circuitry.

The safe-area curves associated with a transistor allow an increase in the power capabilities for pulse conditions when compared to the continuous operating conditions. This results because of the reduced junction temperature that occurs during pulsed-dissipation conditions.

Dynamic safe-area protection still makes use of circuitry to limit the maximum peak current that can flow, and over-voltage shutdown circuitry is also used that causes shutdown to occur as BV_{CEO} is approached to establish the boundaries of the safe-area curve that apply for the particular transistor being protected.

The regions of maximum power dissipation in a power transistor occur where the collector current crosses the base-collector junction. With the predominately vertical transistor action that exists within the IC NPN transistor, this maximum power dissipation region is along the edges of the emitter that face the places where the base contact is made.

The novel peak temperature limiting circuit (that provides for the dynamic safe-area protection) makes use of a PN junction that is located very close to the internal high-temperature regions of the power transistor. This distributed junction temperature sensor is interwoven throughout the complete power transistor structure and consists of an extra emitter diffusion, the "sense emitter," that is located only 0.4 mil away from the active emitters of the power transistor and is also located in the common base region of the power transistor. This sense emitter is forward biased at a current magnitude that will provide a $V_{BE} = 0$ at the desired limiting temperature. As this V_{BE} starts to go negative (with increasing temperature), an op amp senses this polarity reversal and then takes control of the base drive to the power transistor to limit and regulate the maximum hot-spot temperature, regardless of where it occurs within the power transistor structure.

A peak junction temperature limit of 250°C is used for the power transistor region on the die, and a junction temperature limit of 150°C is used for the low-level circuitry of the power op amp IC. This takes advantage of the thermal gradients across the die surface (which result because the power transistor is hotter than the regions occupied by the low-level circuitry) that normally exist in power ICs and eases the requirements for high-temperature operation of the low-level circuitry. If minimum heat sinking of the op amp were used, the package temperature would approach the junction temperature of the power transistor and this would limit the power transistor to a junction temperature closer to the 150°C limit. If good heat sinking is provided, the junction temperature of the power transistor will rise above that of the low-level circuitry (by a maximum of 100°C) and increased output power will be provided.

Dynamic safe-area protection is not limited to ICs. The same distributed sense emitter can be provided for a discrete power transistor, and thermal limiting would be provided by external circuitry.

THE LOW V_{IN}-V_{OUT} PNP VOLTAGE REGULATORS. IC voltage regulators with low dropout voltages (the minimum amount that the input voltage must exceed the output voltage) were originally designed for automotive electronic systems. Low battery voltages result on cold days when the automotive engine is being cranked during starting. This required an electronic voltage regulator that could maintain regulation with the unregulated input voltage only a few tenths of a volt larger than the magnitude of the regulated output voltage. To provide this feature, special IC regulators were designed that made use of a power lateral-PNP transistor as shown in Figure 1-23.

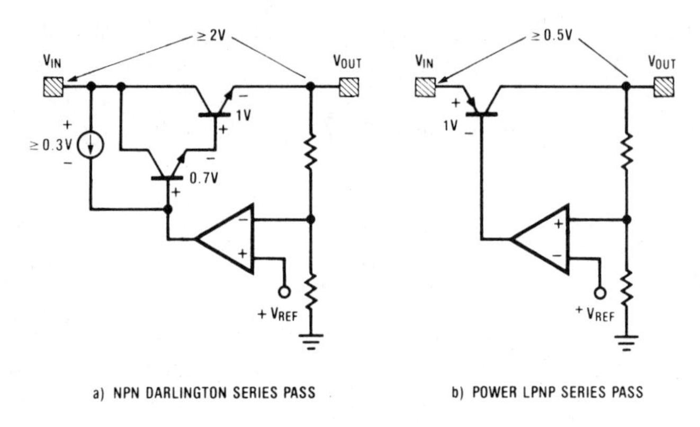

Fig. 1-23 Power LPNP Reduces V_{IN}-V_{OUT} of a Voltage Regulator

Notice that the circuit of Figure 1-23b places the relatively large base-emitter drop of the LPNP power transistor between the unregulated input voltage and ground—not the regulated output voltage. This eliminates the voltage build-up of the typical IC voltage regulator because of the NPN Darlington-connected power transistor, shown in Figure 1-23a.

To handle large currents, a special deep-diffused lateral PNP structure, as shown in Figure 1-24, is made part of the IC voltage regulator.

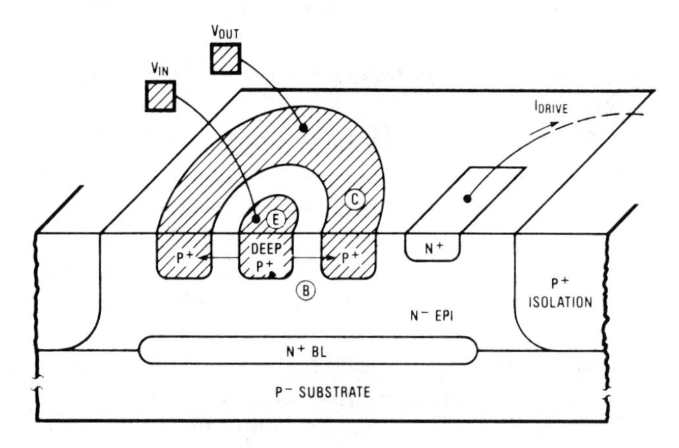

Fig. 1-24 The Power Lateral-PNP Structure

The goal is to have at least a current gain (β) of 10 at the maximum value of collector (output) current. The dependence of current gain on collector current of this LPNP is compared with an NPN and a standard LPNP (all of the same size) in Figure 1-25. Notice the 10:1 increase in collector current that can be carried by the deep base LPNP structure.

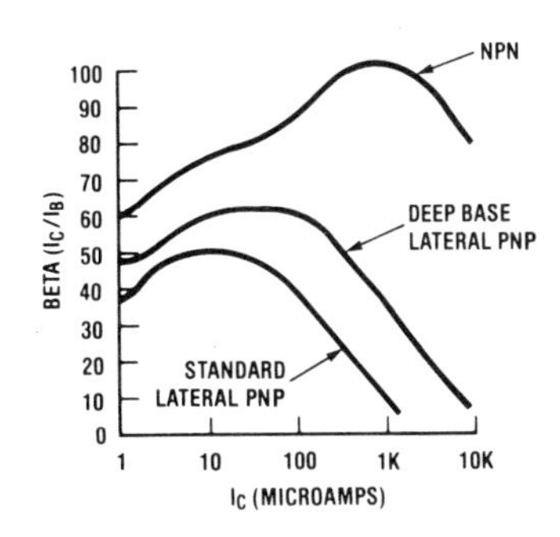

Fig. 1-25 Current Gain Dependence on Collector Current for the Power LPNP

In addition to providing a low V_{IN}-V_{OUT} regulator, the LPNP can allow higher voltages than the standard IC NPN transistor. A special circuit design is used in these voltage regulators that connects only emitters of LPNP transistors to the unregulated V_{IN} line. This results in IC voltage regulators that will not be destroyed by very large positive and negative voltages on the V_{IN} line.

The output current capability of the first of these regulators was 100 mA and products are now available that can carry 1.5A. Similarly, the input transient protection has increased from +40V, −6V to +60V, −30V and the goal for the current designs is ±120V to hold up under the severe voltages that exist on the wiring in automotive electronics.

A COMPATIBLE PNP PROCESS. The continuing need for a low-cost, good, bipolar PNP transistor has finally been solved by adding to the processing complexity of linear ICs. Compatible PNP transistors that make use of dielectric isolation, have been available for a number of years, but these costly processes have not been popular with the large-volume suppliers of linear ICs.

A way of providing a "vertically-integrated PNP" (called, by some manufacturers, a "VIP" process) is shown in Figure 1-26. Many processing changes have been made to create this PNP. A rather deep N^- "well" is located under the PNP transistors, and a P^+ buried layer is located within this well. This buried layer is used to reduce the resistance in the collector bulk regions of the PNP transistors. It provides the same benefits as the N^+ buried layer that is used for the NPN transistors.

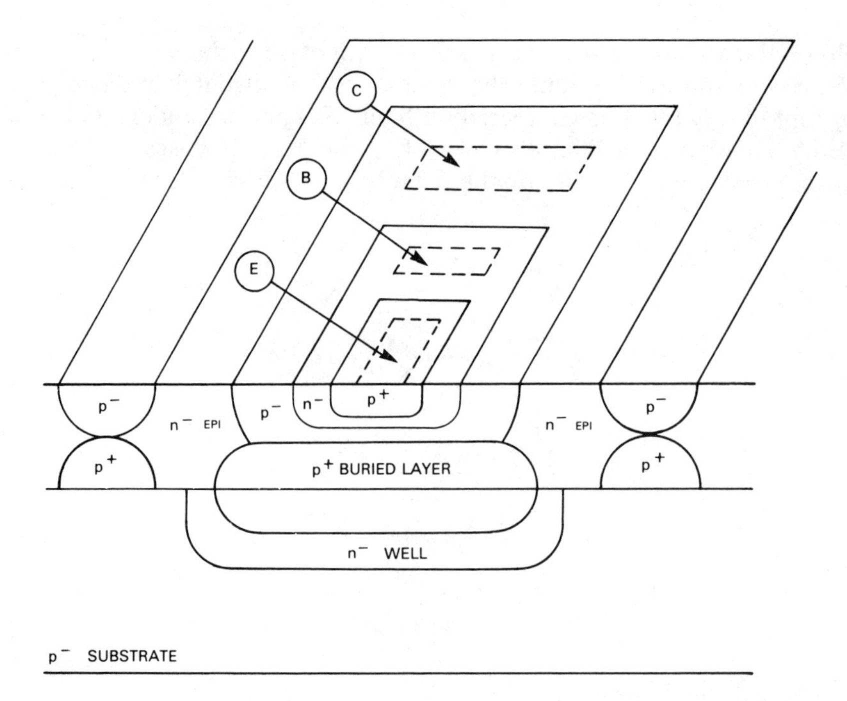

Fig. 1-26 A Structure that Is Used to Form a Compatible PNP Transistor

The epi isolation has been changed to include diffusions that move both down from the surface and up from the bottom through the epi layer (this lower diffusion also provides the buried layer), the "iso up" process. The upper isolation diffusion is also used to form the high-resistivity P-type collector regions for the PNP transistors. This collector diffusion is driven into the P^+ buried layer.

A special N-type base region is then formed within the P-type collector, and a P^+ region, located within this base region, forms the P-type emitter. A contact (for simplicity, not shown on the figure) to the epi layer is used to provide a positive reverse-biasing voltage to insure that the collectors of the PNP transistors never forward bias into the epi region. In addition, a P-type guard ring (also not shown on this figure) is added at the surface, around the perimeter of the P^- collector region. The addition of this compatible, high-performance PNP transistor to the linear designer's bag of tricks has provided, and will continue to provide, many new IC op amps and other products with greatly improved high-frequency capabilities.

AN IN-PACKAGE-ADJUSTABLE, ANALOG TRIMMING RESISTOR. With all of the focus on VLSI, it came as a surprise when three linear IC design engineers (R. L. Vyne, W. F. Davis, and D. M. Susak) won the 1987 Solid State Circuits Conference "Best Paper Award." This is another example of the cleverness and innovativeness of linear IC designers as they make use of "anything at hand" to solve their analog circuit problems.

It has been known that an IC zener diode could be pulsed with a relatively large magnitude of reverse current to cause the diode to become a short circuit. The unusual thing is that a very small "finger" of aluminum/silicon alloy would result, under the surface oxide layer, that would reach across the PN junction and cause this short. This is known as the "zener zapping" trim technique and has been used to trim or adjust the offset voltage of IC op amps to near zero during the wafer sort (final wafer testing) operation. A problem with this technique is the relatively large die area that is needed for a tapped resistor and the many zeners that are placed across the taps so that selected sections of the resistor can be shorted out and thereby adjust the offset voltage to only a few microvolts. Unfortunately, the offset voltage of an op amp shifts during the high surface stresses of scribe, break, and die attach, so the packaged op amps usually have an undesirably higher offset voltage. This zener zapping will also undesirably short out the base-emitter junction of an unprotected input transistor if it is subjected to an electrostatic discharge.

Another solution to this analog trimming problem is presented in R. L. Vyne's U.S. Patent #4,606,781, "Method for resistor trimming by metal migration," dated August 19, 1986. The new resistive trim element (the "trimistor") makes use of the N^+ diffused resistor that is available in the standard linear IC process. With the trimistor, the ohmic contacts are designed to maximize the current density (and therefore the temperature) in the center of the resistor body by making use of tapered contacting regions. When "trimming pulses" of relatively large current (approximately 500 mA for 3 ms) are passed through this resistor, the Al/Si eutectic temperature is reached in these tapered contact regions at the surface of the resistor and a small amount of the silicon that was originally dissolved in the Al/Si alloy of the ohmic contact moves into the aluminum surface metal (in the direction of the electron flow). The contact regions adjust to this loss of silicon by dissolving more silicon from the small hot regions at the surface of the diffused resistor. This forms a sliver of Al/Si alloy that is approximately 3 μm wide and 1.5 μm deep (with a cross-sectional area of 3.5 μm^2) that moves in small steps, one for each trimming pulse, down the resistor body (in a direction opposite to the electron flow). In this manner an 80-ohm resistor can be incrementally reduced in value by 0.1-ohm steps.

A diagram of the trimistor is shown in Figure 1-27. For clarity, the aluminum interconnect metal is not shown so that the details of the wedge-shaped contact regions can be displayed. In addition, only one extra contact region is shown between the end contacts to the resistor. In practice, many additional contacting regions (a total of 10), including their small isolated patches of aluminum metal, were spaced down the resistor body

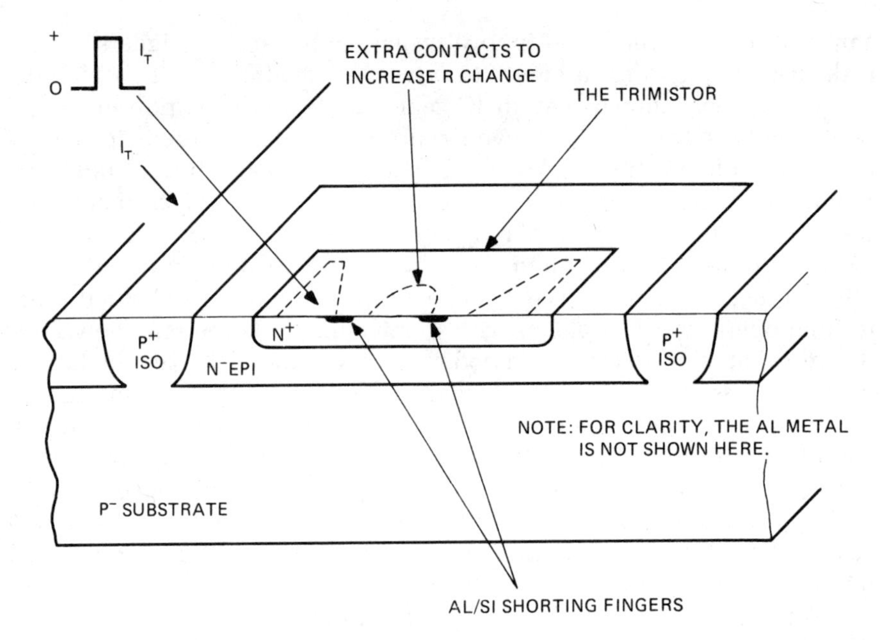

Fig. 1-27 The Trimistor

to magnify the effects of the trim on the total value of the resistance change. The darkened regions at the edges of the contact windows in this figure are used to represent the shorting slivers or fingers of Al/Si alloy that propagate down the surface of the resistor in the direction of the trim current pulses. Notice that the end contact, on the right-hand side where the trim current leaves the resistor, does not have any finger growth associated with it.

This trimistor was used to allow simply placing JFET, unity voltage-gain, source followers (Q3 and Q4) in front of a bipolar op amp, as shown in Figure 1-28. The benefits of essentially no input current are therefore provided in this modified op amp because the external inputs are buffered from the relatively high input current of the bipolar op amp by the high dc input impedance of the JFET source followers. The JFET biasing current sources Q1 and Q2 have trimistors R1 and R2, respectively, in series with their source leads. The biasing current magnitude will therefore increase as the trimistor is reduced in value during the trim operation. This change of bias current flow through Q3 and/or Q4 is used to alter the dc gate-source biasing voltage of these source followers and thereby adjust the offset voltage of the overall op amp.

To eliminate any stress-related shifts in the trimmed offset voltage, the trimistor should be accessable for trimming in the packaged IC. In the past, this has required additional package pins to allow accessing the on-chip trimming circuitry. In this op amp, the existing pins were also used for in-package trim by making use of the circuit shown in Figure 1-29

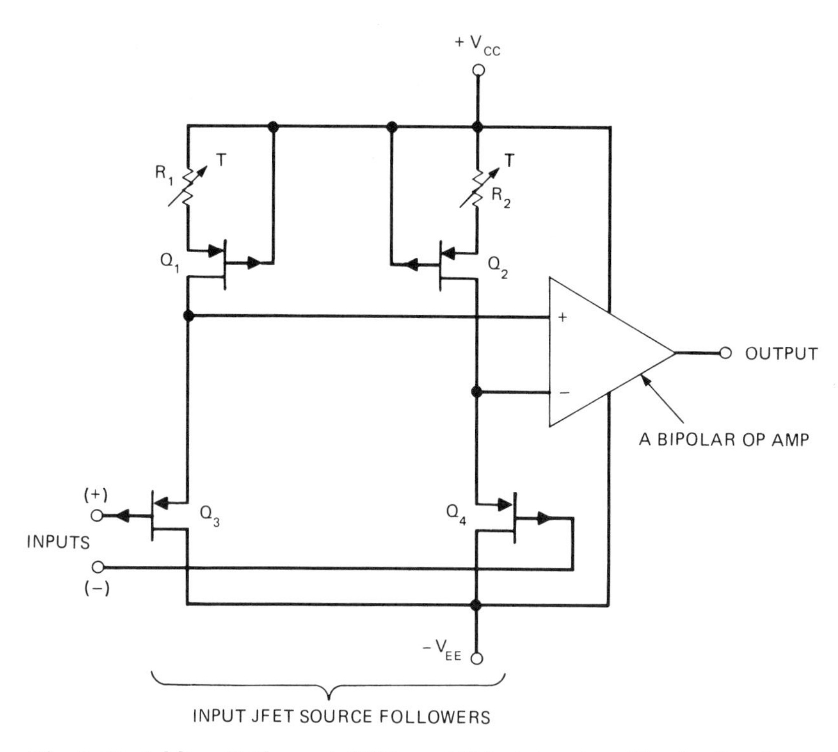

Fig. 1-28 Adding P-Channel JFETs and Trimistors to a Bipolar Op Amp

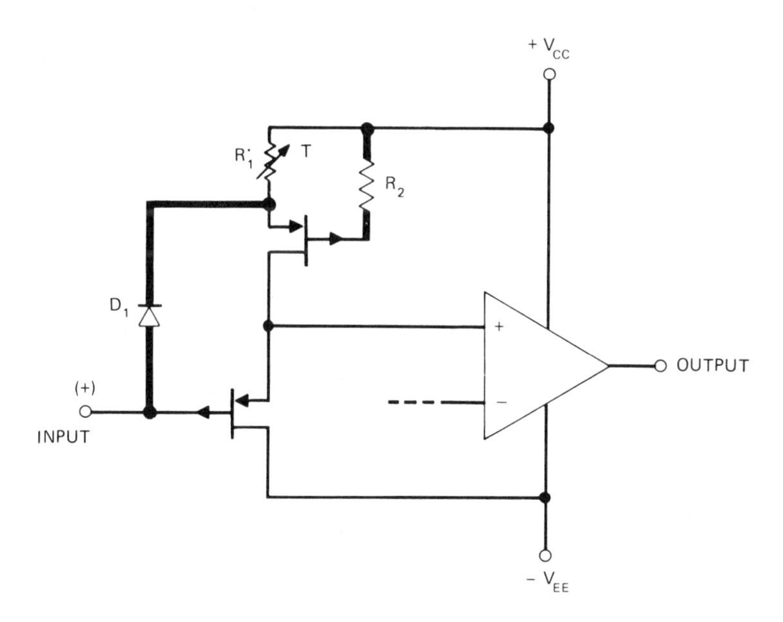

Fig. 1-29 Components Added for In-Package Trimming

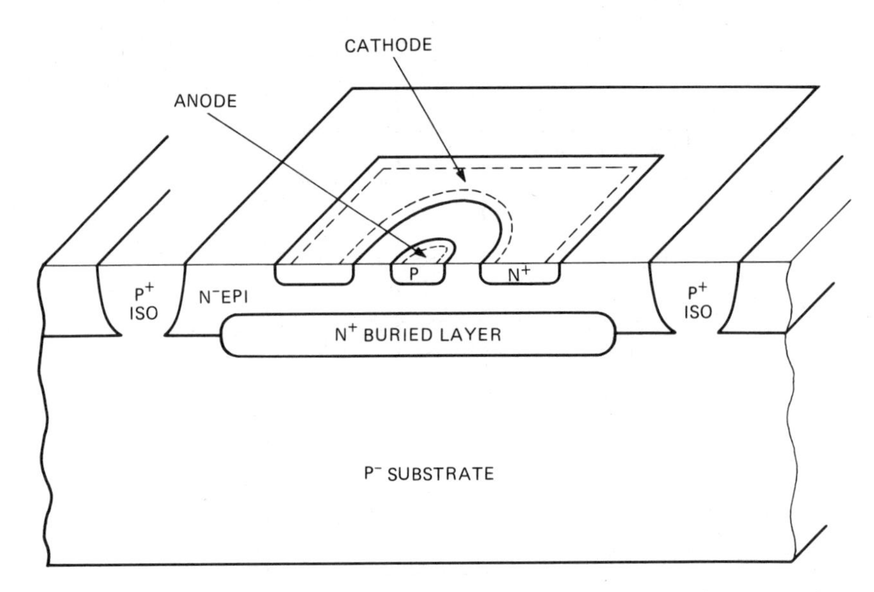

Fig. 1-30 The High-Voltage, High-Current, Low-Leakage Isolation Diode

where a diode (D1) and a resistor (R2) were added to each of the input JFET source followers. During trimming, the source of the JFET current source will be driven quite positive, and the resistor R2 (2.2 k) limits the current that flows through the forward-biased source-gate junction of the current-source JFET. Now, the trim current can enter the input pin, go through the normally isolating diode (D1), pass through the trimistor (R1) to affect the trim, and exit via the positive power supply pin ($+V_{CC}$). In normal op amp applications, the diode D1 is reverse biased and therefore disconnects the input pin from the trimistor.

Unfortunately, the reverse-biased leakage current of this isolation diode flows through the input leads of the op amp. A special diode was therefore needed that had a high breakdown voltage, low reverse leakage current and capacitance (and therefore also a small die area), and could handle the large trimming current pulses. These characteristics were simultaneously achieved by using the minimal-area diode structure shown in Figure 1-30. The normal base diffusion forms the P region and the N^- epi is the N region of this special diode. The forward voltage drop at 500 mA is 1.7V, the reverse breakdown is 125V, and the leakage current is negligible compard to the JFET gate current.

The use of these new trimistors allows initial offset voltages as large as 7.5 mV to be in-package trimmed to 10μV. The next question is, What other useful applications will the linear IC designers find for this new circuit component? (For more information see: "A Monolithic P-Channel JFET Quad Op Amp with In-Package Trim and Enhanced Gain-Bandwidth Product," R. L. Vyne, W. F. Davis, and D. M. Susak, *IEEE Journal of Solid-State Circuits*, vol. SC-22, no. 6, December, 1987.)

CHAPTER II

The Move to MOS

The move to MOS ICs was not easily accomplished. The reliability problem of the early PMOS products was a big obstacle for the IC manufacturers to overcome. The IC designers also had to become familiar with the different operation of an MOS device. With MOS logic, there is no dc current flow at the inputs, so fan out is limited only by the minimal acceptable rise and fall times that result from the increased capacitive loading of the additional MOS inputs and the increased stray wiring capacitance that also accumulates from the large number of inputs.

2.1 COMPARING TRANSISTOR PERFORMANCE - MOS VERSUS BIPOLAR

The internal mechanisms that take place within a bipolar transistor are more complex than those of a MOS transistor. Even the name *bipolar* was chosen because both *majority and minority* charge carriers are involved. An MOS transistor is called a *majority carrier* transistor because only *majority* carriers are involved in the basic operation of this device.

A two-dimensional representation of a bipolar transistor, operating in the forward active region, is shown in Figure 2-1. We will quickly review this bipolar transistor action so we can point out the differences that exist in MOS transistor operation. If we neglect charge recombination at the surface of the emitter-base junction, there are two principle mechanisms taking place: electron charge injection from the emitter into the base region and the transport of this charge across the width of the base region over to the edge of the collector-base space charge layer. At this edge, the charge is collected and results in a current in the external collector lead.

When the emitter-base diode of the transistor shown in Figure 2-1 is forward biased, both sides of this diode are injecting carriers. This *back injection of holes* by the base into the emitter region reduces the current gain because it constitutes an undesired component of base current flow, a high-β transistor has only a very small base current. The impurity doping density of the emitter is increased over the impurity doping of the base to reduce this back injection component.

The sum of the collector and base currents must equal the emitter current. Further, the ratio of the collector current to the base current is

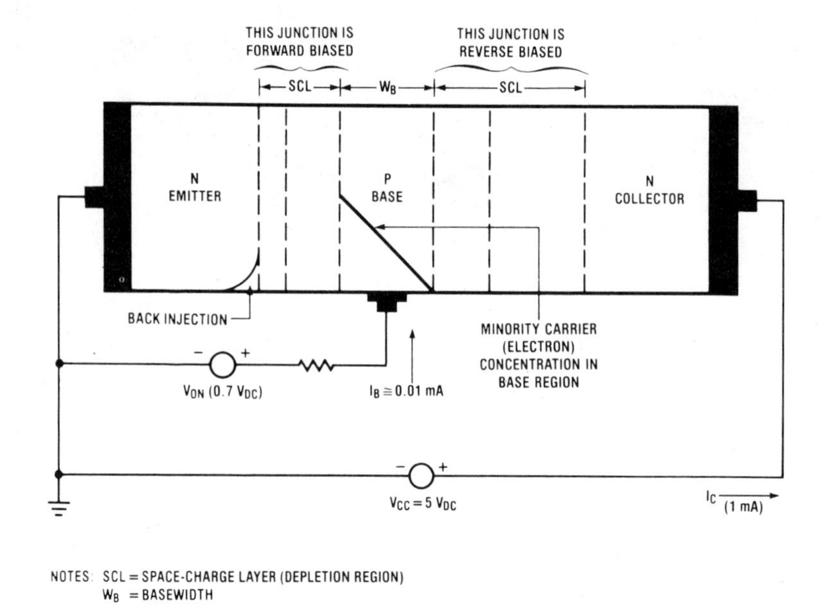

Fig. 2-1 Bipolar Transistor Representation in the Forward Active Region

the current gain of the transistor, called *beta* (β), and large values of beta are desired.

Most of the carriers that are injected by the emitter into the base region find their way to the edge of the collector-base space charge layer. The *base transport factor* is a measure of how many carriers (electrons) make it across the basewidth without recombining along the way with the majority carriers (holes) that exist within the P-type base region. Transistors with relatively narrow basewidths and light doping of the base region generally have larger values of current gain (I_C/I_B) because, for these devices, the base transport factor is close to unity.

In contrast with a bipolar transistor, an MOS transistor consists of relatively heavily doped source and drain regions, of the same impurity doping type, that are located within a lightly doped *body* or substrate region of the opposite impurity doping type, as shown in the NMOS transistor structure of Figure 2-2. An enhancement-mode MOS transistor will be in a nonconducting state when the gate-to-source voltage is zero volts. As shown in this figure, a reversed-biased diode then isolates the drain from the substrate.

When a voltage that exceeds the *threshold voltage* of this transistor is applied to the gate, sufficient electron charge is extracted from the source and held at the silicon surface, under the oxide layer, in the channel region. The channel region exists at the surface of the silicon between

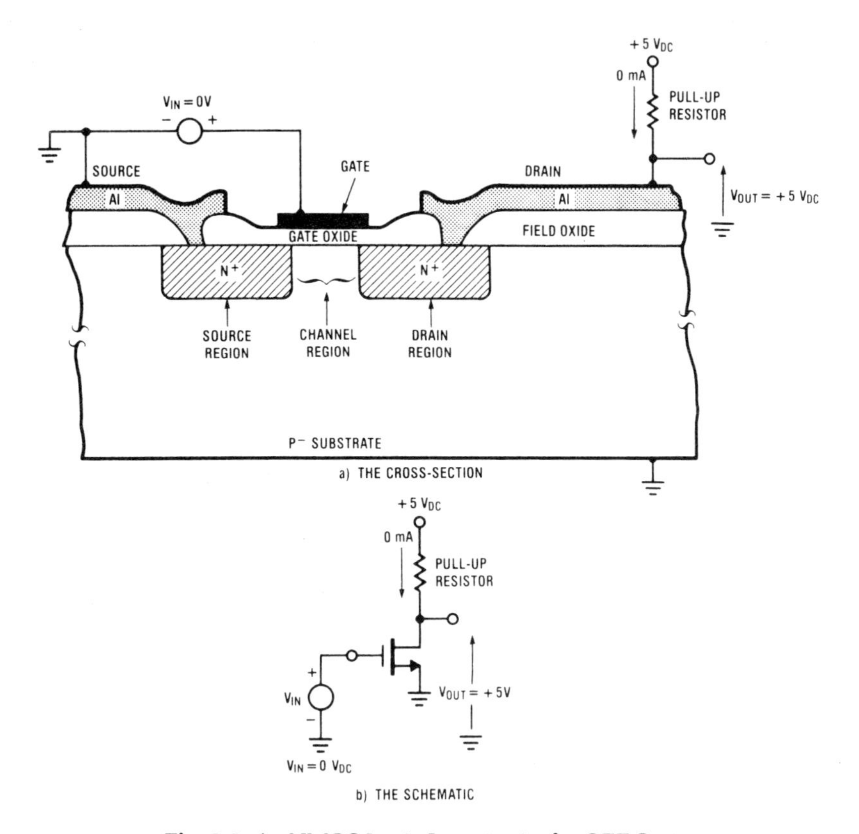

Fig. 2-2 An NMOS Logic Inverter in the OFF State

the source and drain regions. This causes the apparent doping type of the channel to change. Now there are no longer isolating diodes separating the source and drain regions. The same doping type extends from the source to the drain and therefore current can flow. The transistor can now conduct current from the source to the drain; it is ON, as shown in Figure 2-3.

Larger applied gate-source voltages increase the conductivity of the channel and thereby increase the source to drain current flow. Gate voltages less than the threshold voltage do not form a channel or form only a weak channel, so there is little or no current flow from source to drain.

The MOS transistor is often considered to be a voltage variable resistor. The applied gate voltage controls the source to drain resistance. This view is especially useful when the MOS device is serving as an electrically actuated ON-OFF switch, the analog switch. Notice that there is also no gate current flow, therefore all of the current that is injected by the

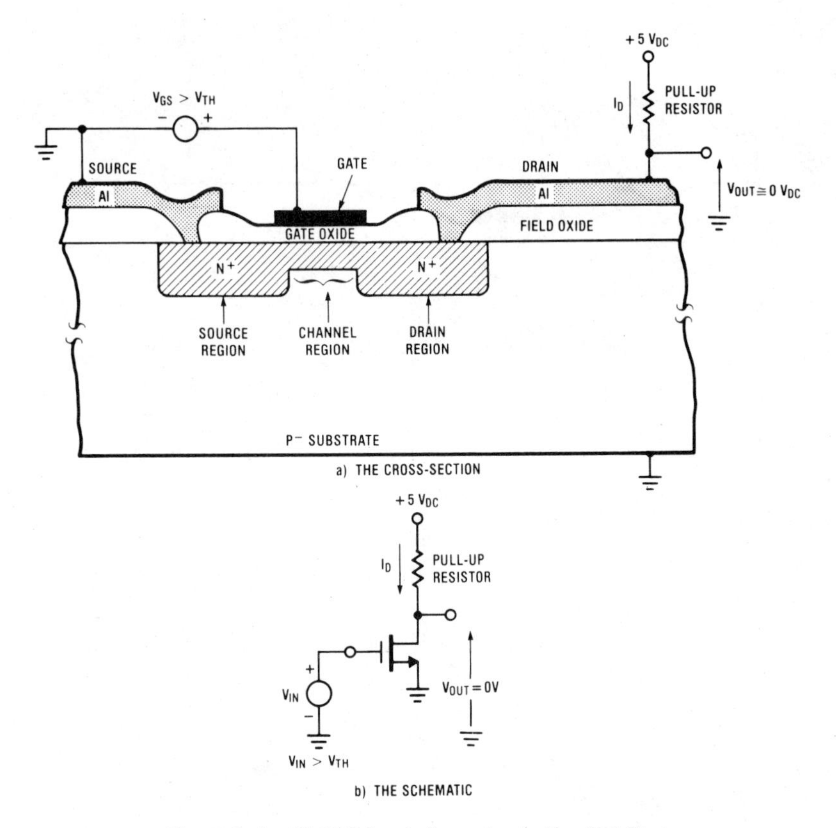

Fig. 2-3 An NMOS Logic Inverter in the ON State

source into the channel appears in the drain lead. In many applications, one or the other of these benefits of the MOS transistor over a bipolar transistor can be important in a circuit.

An MOS transistor can be modified by making use of an ion implanter. Ion implantation is often used to place a very light impurity doping directly in the surface of the channel region. This implanted impurity doping creates a *built-in channel* between the source and drain regions, even when the gate-source voltage is zero. This new device is called a *depletion mode* MOS transistor because the application of a reverse gate-source voltage is required to *deplete* or reduce the channel formation and thereby reduce the drain-source current flow. Depletion mode FETs are used to replace a pull-up resistor in a logic gate. These devices have the FET *pinchoff* benefit of maintaining a more constant current flow, when compared with a resistor pull-up, as the output voltage of the gate rises toward the supply voltage level. The voltage-current characteristics of a depletion mode pull-up, a resistor pull-up, and an ideal current-source pull-up are shown in Figure 2-4.

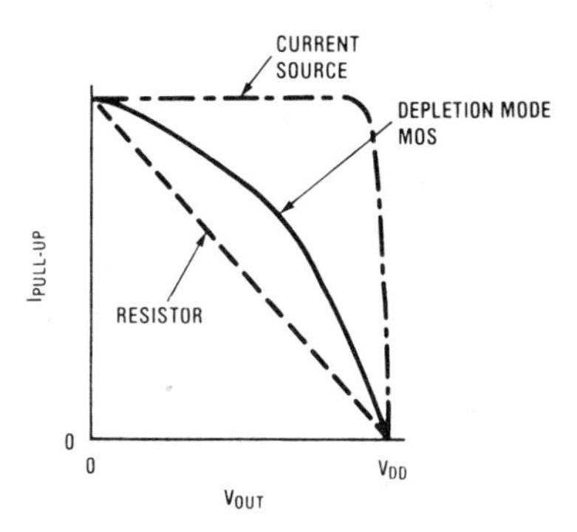

Fig. 2-4 Voltage-Current Relationship of Various Logic Pull-Up Devices

It is the dc current flow through the pull-up device, whether a resistor or a depletion-mode FET, when the output transistor is ON that causes dc current drain in bipolar, PMOS, and NMOS logic circuits. The usual assumption is that one-half of all of the logic circuits in a system will be in the dissipating ON state. As we will see later in this chapter, the novel *gated load* of CMOS logic circuits eliminates this static dc power drain.

Effects of Layout Geometry on Performance

With all of the emphasis on obtaining very small feature sizes in IC fabrication, it may not be clear why all of the transistors that are used on an IC die are not designed to be of the smallest possible physical size. Transistor sizes are increased to handle larger values of current. Bipolar transistors have several high current mechanisms, such as reduced emitter efficiency and basewidth stretching, that degrade the performance of the transistor at large current densities. The important thing is therefore to keep the *current density* within a transistor structure at a low enough value that these large signal problems do not degrade performance.

MOS transistors are enlarged to deliver more current, making the capacitively loaded nodes driven by this transistor change voltage more rapidly, or to reduce the value of the resistance that exists between source and drain when the transistor is switched ON.

The disadvantages of all large-geometry transistors are an increase in the chip size and an increase in the parasitic capacitances associated with the larger transistor. This capacitance increase also raises the current drain of the IC during continuous switching and requires that the driver transistors also be made larger in size so that they can provide the larger currents necessary to rapidly charge the large parasitic capacitances associated with the main power or output transistor.

Transconductance

As mentioned earlier, large values of transconductance (g_m) are desirable and result when the control element is in closest proximity to the device structure. In MOS devices, this requires having the gate as close as possible to the channel. Early MOS processes used relatively thick layers of gate oxide (1,200 to 1,500 Å) and this reduced the g_m of these transistors. Modern processes have steadily reduced the thickness of the gate oxide layer from 400 Å to less than 200 Å, which has increased the g_m. Therefore, there is an increase in the amount of current that these transistors can supply with a given gate-source overdrive voltage.

In MOS transistors, g_m also directly depends on the carrier mobility. This causes N-channel MOS transistors to have approximately three times larger g_m than P-channel transistors. To achieve symmetrical rise and fall times when both transistor types are used in a logic circuit, as in complementary MOS (CMOS) logic designs, the P-channel transistors are usually designed to have three times the width of the N-channel transistors, because the output current of an MOS transistor also depends on the geometry of the transistor, as shown in Figure 2-5. Think of it as three transistors in parallel. Devices with a large ratio of channel width to length, W/L (also called Z/L), therefore conduct correspondingly larger values of drain current.

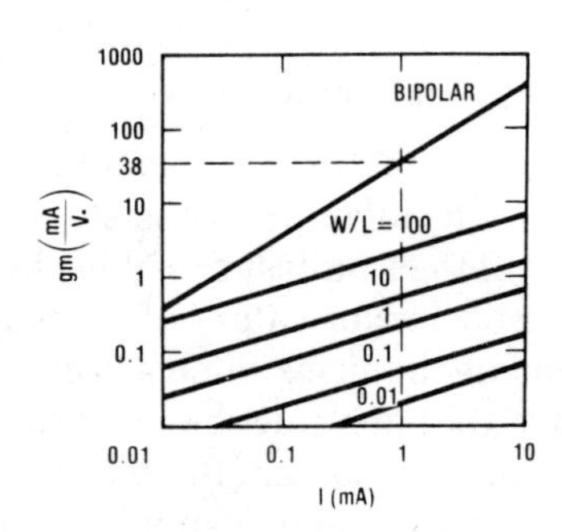

Fig. 2-5 Comparing the Transconductance of a MOS Transistor with a Bipolar Transistor

It would appear, from this figure, that a MOS transistor with a large enough W/L would have a higher g_m at a particular biasing current than that of a bipolar transistor. What is missing in this diagram is the fact that a transistor with such a large W/L, when operated at the low values of current that are indicated on this figure, would be in the *sub-threshold* region of operation. MOS transistors, when operated in the sub-threshold region, have lower g_m than is indicated on this figure.

The transconductance of a bipolar transistor is given by

$$g_m = \frac{1}{r_e} = \frac{I_E}{V_T}$$

where: r_e is the small-signal emitter resistance
I_E is the emitter dc biasing current
$V_T = \frac{kT}{q}$ (= 26 mV at room temperature)

and

k = Boltzmann's constant (1.38×10^{-23} joules per °K
q = the charge of an electron (1.6×10^{-19} coulombs)
T = absolute temperature in degrees Kelvin

$$T(°K) = T(°C) + 273$$

The transconductance of an MOS transistor in saturation (that will be defined shortly) can be expressed as

$$g_m = \frac{1}{r_s} = \frac{2I_D}{\Delta V}$$

where: r_s is the small-signal source resistance
I_D is the drain biasing current

and

$$\Delta V = V_{GS} - V_{TH}$$

where: V_{GS} is the gate-source ON voltage

and

V_{TH} is the threshold voltage of the MOS transistor

(This useful expression is similar in form to the equation used for a bipolar transistor and was first derived by Bill Jett.)

Another useful set of equations is:

$$g_m\ (\text{N-ch}) = \frac{(W/L)\ \Delta V}{40}\ \text{millimhos}$$

$$g_m\ (\text{P-Ch}) = \frac{(W/L)\ \Delta V}{150}\ \text{millimhos}$$

where the numbers in the denominators are the respective "g_m factors" of the MOS transistors.

All of these expressions for g_m apply for the MOS transistor when operated in the *saturated or current source region*, which exists when

$$\left|V_{DS}\right| \geq \Delta V$$

In this operating mode of the transistor, the drain current remains essentially constant even though the drain voltage may change. The high value of output impedance and the large allowed range of drain voltage that exists when the MOS transistor is in the saturated mode of operation is most useful for amplifier applications.

For smaller drain-source voltages, the MOS transistor operation is said to be in the *linear or resistive region* because the drain-source current increases as the drain-source voltage increases. This represents a low value of output impedance and is therefore not desirable for amplifier applications. Logic circuits operate the MOS transistors OFF and ON. When ON, the transistors are in the linear mode of operation. The emphasis on logic design has increased the MOS transistor modeling effort for the linear mode. Modeling of the MOS transistor in the saturated mode of operation is not as well-developed.

Frequency Response

The frequency response of an amplifying device can be increased by reducing the path length of the current carriers. This implies a narrow basewidth for a bipolar transistor or a short channel length for an MOS transistor. In either transistor, when this path length is made very short, the transistor will no longer support high voltages on the output terminal, either collector or drain. A *punch-through* breakdown voltage limitation exists, which is the same basic physical mechanism for either device. This is the reason for the reduced operating voltages that must be used for the advanced MOS products. The standard 5-volt supply has to be reduced to 2 or 3 volts, and the ultimate processes of the future may require less than a 1-volt power supply.

Junction capacitances are common to both the bipolar and the MOS transistors. The magnitudes of these parasitic capacitances directly depend on device size and undesirably limit the frequency performance. Newer processes eliminate as much of the parasistic PN junction area as possible. This motivated the development of a silicon on sapphire CMOS process and also created interest in *oxide isolation* techniques for both MOS and bipolar transistors, where the PN junctions have the vertical edges replaced with a layer of SiO_2. We will have more to say about this in the next chapter.

Output Impedance

An ideal amplifying device has an infinite output impedance. Actual amplifying devices have a finite output impedance because the output current undesirably increases as the magnitude of the output voltage is increased. In both the bipolar and MOS transistors, the basic reason for a finite output impedance is the same; changes in the output voltage change the amount of the depletion layer spread within the silicon and this modifies the effective length of the path the charge carriers must traverse from their source to the eventual output terminal. In the bipolar transistor, this is the *Early effect* or *basewidth modulation* phenomenon. In the MOSFET, this is called *channel length modulation*. Both of these effects become more significant as the initial values of the path lengths are made shorter, as happens in smaller devices. In general, the low output impedance effect is more of a problem with the MOSFET and tends to limit the amount of voltage gain that can be obtained in a single MOS-transistor amplifying stage.

Modeling Accuracy

The accuracy that is needed in the computer modeling of a transistor for use in circuit analysis programs depends on the intended usage and the accuracy needed in the final result. Unfortunately, as more elements and physical effects are accounted for in each individual transistor, the total number of transistors that a computer can handle in a given circuit analysis becomes limited.

The largest expenditures of modeling resources has been in support of dynamic random access memory (DRAM) products. In most digital circuits, the main interest of the circuit designer is to predict the time response. This allows many modeling simplifications to be used when compared, for example, with the complexities of linear circuit design. The details of many of the problems associated with linear sampled data systems, such as charge loss in the signal transfer through analog switches and the effects of clock-signal charge-coupling through analog switches, are generally not handled with sufficient accuracy to predict the effect on the overall circuit performance.

A great saving in computer run time can be made by using *macromodels*. These model only the overall transfer function of a more complex circuit block, such as an operational amplifier (op amp) or a logic building block. Larger circuit blocks of a system can be handled on a computer by making use of this simplification.

In general, the costs of computation and memory are declining so rapidly that large computer and memory resources can be assigned to an individual designers' workstation. Extensive computer simulation and even automatic IC mask generation are necessary to handle the complexities of the new VLSI chips.

Temperature Effects

Temperature effects on circuit performance depend upon the active devices that are used in the design. For example, raising the ambient temperature of an MOS device will reduce the g_m, and therefore the output current, at the rate of approximately $-0.3\%/°C$. The response time of MOS products degrades at this same rate as the ambient temperature is increased. This speed degradation with temperature limits the performance of complex NMOS products because their relatively large power dissipation heats up the chip. We will see that a reduction in power dissipation is one of the main reasons for the shift to CMOS in the present VLSI era. Further, the response time of MOS products typically degrades as the power supply voltage is reduced, at the rate of approximately a 3.5% increase in propagation delay per 100 mV of power supply voltage reduction.

MOS products generally work better (faster) when cold, which is just the opposite of the performance of bipolar products. For a bipolar transistor, g_m also decreases with increasing temperature, but the current gain (beta) increases at approximately $1\%/°C$, which is the principle reason for the increased base current that can degrade circuit performance at reduced temperatures. Problems do occur in bipolar saturated logic because of the effects of temperature on the collector saturation voltage and the base-emitter ON voltage. Both of these go the *wrong way* as temperature is increased and the net result is that the noise margin between the logic voltage levels is reduced. V_{BE} falls at approximately -2 mV/°C and the collector saturation voltage in standard TTL circuits increases at approximately $+0.5$ mV/°C.

High Power Limits

The high power limits of ICs relate to four dominant effects: the transistors must be large enough to carry the required currents, the chip must be heat sunk to limit the maximum junction temperature to approximately 150°C, the currents that flow must not cause the surface metal or the bonding wires to migrate or to melt, and a bipolar power transistor must not be allowed to enter a destructive breakdown mode.

MOSFETs have natural characteristics that make them well-suited to high power applications. This fact has been exploited mostly with discrete power FETs that are now in competition with bipolar power transistors. Linear power MOS IC products are now starting to appear. We will have more to say about this later in this chapter when the DMOS device is introduced. As MOS analog circuits become more common, it can be expected that power MOS ICs will strongly compete with bipolar power ICs.

2.2 PMOS, THE FIRST LSI

Although the idea for an MOS transistor predated the bipolar transistor, the MOS device was much more difficult to fabricate. In contrast to the relatively heavily doped silicon regions of the low voltage T^2L products, MOS processes require a very lightly doped region in which to build the transistors. These lightly doped regions could be easily inverted by the very small amounts of undesired impurity doping atoms that contaminated the early fabrication lines.

One of the more common problems was sodium contamination. It has been estimated that there are enough sodium atoms in the salt (sodium chloride) residue of one fingerprint that, if placed in critical areas, could contaminate every MOS wafer ever fabricated! This "people grease" problem has shut down wafer fab lines when proper precautions were not followed. The sodium ion is very mobile and easily moves around within the SiO_2 layers. Raising the temperature increases the thermal energy of these sodium ions and increases their movement. Further, an electric field will also make these mobile positively-charged ions cluster at, or near, if an insulating layer isolates the direct contact, the negative electrode. The increased cleanliness needed for MOS IC production required special procedures to be instigated on the MOS fabrication lines.

The first of the MOS processes was *P-channel MOS* or PMOS. The choice of PMOS rather than an N-channel process (NMOS) was based on the different way that sodium ion contamination in the gate oxide alters the threshold voltage of an MOS transistor. The threshold voltage of a sodium-contaminated N-channel transistor is reduced in magnitude as indicated in Figure 2-6. But the threshold voltage of a sodium contaminated P-channel transistor is increased, as shown in Figure 2-7.

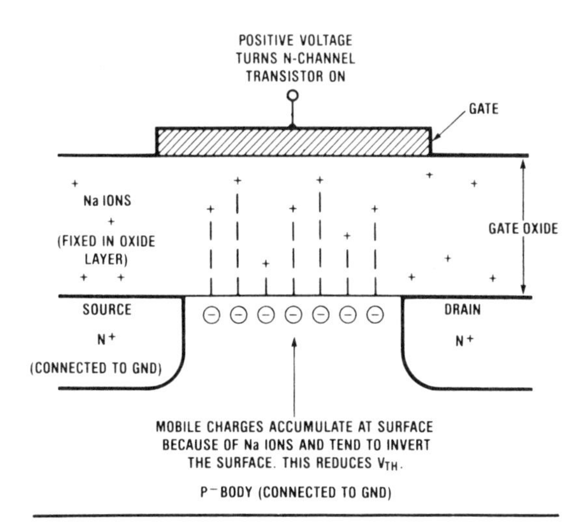

Fig. 2-6 Sodium Contamination in an N-Channel Transistor

89-319

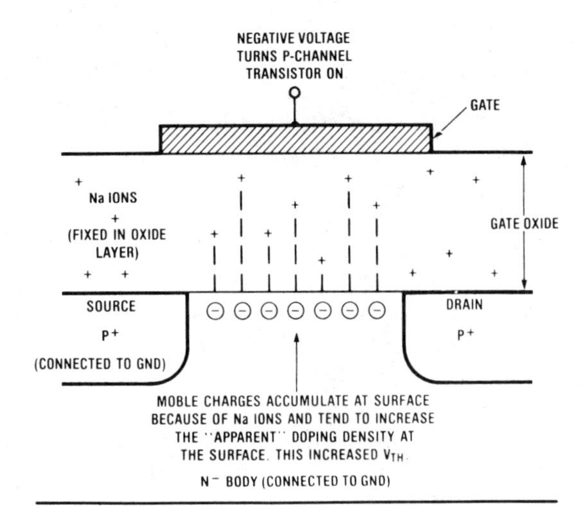

Fig. 2-7 Sodium Contamination in a P-Channel Transistor

For example, if a P-channel transistor is fabricated with an initial threshold voltage of 1.5 volts, sodium contamination would raise this threshold voltage to perhaps 3 volts or more. The values of power supply voltage that were used for the early PMOS circuits were −12 and −18 volts. These large voltages could still turn ON a contaminated P-channel transistor, so some degree of sodium contamination could be tolerated and the PMOS logic circuits would still function.

Because an N-channel transistor is affected in the opposite way, sodium contamination causes the threshold voltage to drop in value. A problem exists when the threshold voltage drops to zero volts. This contaminated transistor then cannot be turned OFF. It has been converted from an *enhancement mode* of operation to a *depletion mode* of operation. This sodium contamination problem therefore delayed the introduction of NMOS products.

The early PMOS wafers required fewer processing steps than were needed for bipolar fabrication. This simplicity, in addition to the simpler device structure, allowed acceptable yields for more complex logic chips. Thus, it was PMOS that allowed the *Large Scale Integrated* (LSI) circuit chips (300 to 3000 gates per chip) to be brought to the marketplace in the early to mid-1970s. These early LSI products did not easily interface to the existing bipolar T^2L logic voltage levels and special logic-voltage-level translator circuits were needed at both the input and the output of these PMOS LSI chips.

Although PMOS processes and products brought in the LSI era, today PMOS is essentially obsolete and the few products that are kept in production are generally built on a CMOS line using a simplified process.

A disadvantage of PMOS products is that the charge carriers for this majority or *unipolar transistor* are holes, and the mobility of holes is approximately a factor of three smaller than the mobility of electrons. This mobility difference reduces the transconductance and therefore the speed is reduced, as compared to logic circuits made with the higher performance N-channel transistor.

Solving the Early Reliability Problems

The early problems with sodium contamination of PMOS wafers were solved by a general increase in the cleanliness of the process flow, the *clean oxide* approach, and the use of *getters*. A getter is a substance that can chemically or physically *trap* undesired atoms or molecules and thereby take them out of circulation. Getters were also used inside the glass envelopes of vacuum tubes to trap reactive gas molecules to insure a more perfect vacuum.

In the processing of silicon, it has been found that phosphorus atoms will attract and hold onto sodium ions. These phosphorus atoms also getter other metallic impurities that would otherwise degrade transistor performance. For this reason, phosphorus-doped glasses are used in the manufacture of ICs. However, if the phosphorus concentration exceeds 7% by weight in the SiO_2 film, then the presence of moisture can create *phosphoric acid* which will disastrously etch the aluminum interconnect lines. The lack of control of the phosphorus content in oxide layers has created many IC reliability problems in the past. Some MOS processes employ a phosphorus gettering step and then strip off this gettering oxide layer prior to finishing the wafer processing.

In the early days of PMOS, some companies were proponents of the clean oxide solution, some were dedicated to the use of getters, and the more successful companies made use of both techniques. These modifications to the processes allowed the fabrication of reliable PMOS products and initiated the LSI era. It is not often realized that the early PMOS memory products successfully took computer memory away from the use of magnetic cores, PMOS LSI products initiated the early hand-held calculators, and a PMOS process was used to build the first microprocessor.

Basic PMOS Logic Circuits

To appreciate the reasons for the technological evolution of PMOS processes, we will consider the basic logic inverter circuit shown in Figure 2-8. Notice that the supply voltages are both negative and that the logic voltage levels are approximately 0 volts and −12 volts. This caused most PMOS products to employ *negative logic*—where a logical "1" is represented by the most *negative* voltage level.

The extra −18-volt power supply is used to provide a large overdrive at the gate of the enhancement-mode load device, Q2. This insures that the output voltage swing of the inverter will come very close to the −12-volt power supply voltage level. The inverting transistor, Q1, has to be designed large enough, a sufficient W/L, in comparison to the size of the load device to allow the output voltage to come very close to ground level. This is called a *ratioed design.*

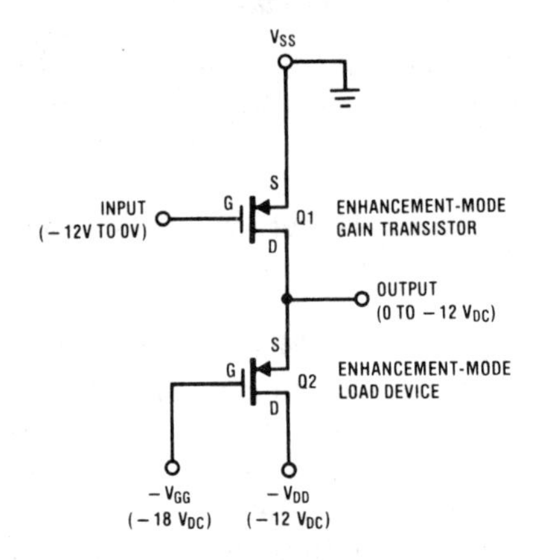

Fig. 2-8 A Basic PMOS Logic Inverter

The Benefits of a Depletion Load

The extra −18-volt power supply was an inconvenience to the system designers. Technologists searched for ways to eliminate the requirement for this extra power supply voltage. This was the motivation that produced the *depletion-mode load device.* This new load element, shown with a special symbol in Figure 2-9, had a channel permanently formed by

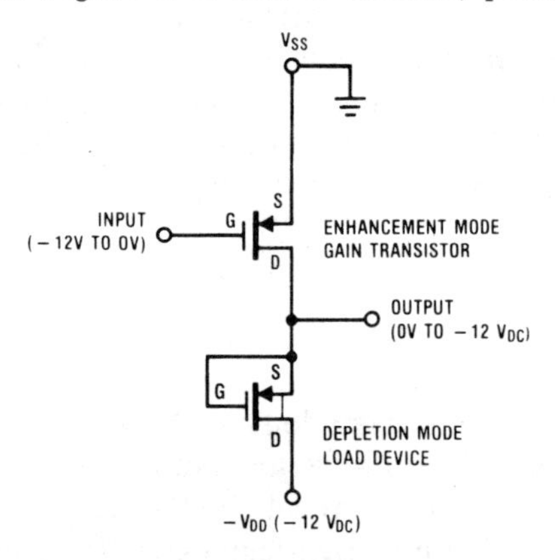

Fig. 2-9 A PMOS Logic Inverter with a Depletion-Mode Load

making use of a low doping concentration ion implantation directly in the channel region. Now current could be supplied by this load device even though the gate-source voltage was operated at 0 volts. This removed the requirement for the extra power supply voltage and still allowed the output voltage to swing very close to the – 12-volt power supply voltage level.

Poly and the Self-Aligned Silicon Gate

A basic alignment problem exists in the processing of MOS circuits. In the early PMOS process, the gate was simply a metallized area, using the aluminum interconnect metal, that covered the channel. The problem was that this metal had to *completely* cover the channel region that exists between the source and drain diffusions. Because these diffusions were located by an *earlier* masking step, the gate metal had to be precisely aligned to this previous masking operation. For example, if a small sliver of the channel was not completely covered with the gate metal, that portion of the channel would not invert under the influence of an ON voltage applied to the gate metal, so this small sliver would interrupt the channel and prevent the transistor from turning ON.

The use of two different masking steps meant that the mask-to-mask alignment would have to be very carefully done. To ease this requirement, the gate metal was simply made larger to insure that the channel region would be covered, even with some amount of mask-to-mask misalignment. This resulting overlap of the gate metal over the drain and source regions increased the input capacitance of the logic circuits and slowed down the response times.

To solve this problem, a way was needed to align the gate to the source and drain regions that did not depend on precise mask-to-mask alignment. The basic idea that was used to solve this problem was to use the *material* of the *gate* to also *serve as the diffusion mask* for the location of the source and drain regions. Aluminum could not be used for such a diffusion mask because it would melt at the high temperatures that were used in the diffusion furnaces. A layer of silicon was therefore formed on the oxide surface of the wafer before the formation of the drain and source regions. This was similar to the epitaxial deposition of silicon onto a *monocrystalline* silicon substrate, except now there was no ordered crystal lattice present in the underlying *amorphous* oxide layer. This caused the added silicon to deposit only in ordered *clumps* and not in a well-ordered *crystal lattice*. Therefore this has been called a *polycrystalline* silicon layer, or simply a *poly* layer. This poly was then masked and etched to define the gate regions of the MOS transistors. These poly gates then served as masks for the impurity diffusion that formed the sources and drains. This was the first use of a poly layer and the interesting thing is that all modern MOS processes still employ at least one poly layer, or the newer metal silicides, to serve as the mask that determines the channel length, the separation of the source and drain regions. This is the *Self-Aligned Silicon Gate* technology.

This poly layer was doped with N-type impurities to reduce the resistance and make it a better conductor. Heavily boron (P-type) doped polysilicon gates create threshold voltage instability problems and therefore N^+ doping is generally now used for all poly gates, whether for an N-channel or a P-channel transistor. The N^+ doping of the poly reduces the threshold voltage of an N-channel transistor and raises the threshold voltage of a P-channel transistor. Direct P-type implants into the P-channel transistor are used in modern CMOS processes to initially create a depletion-mode device that later is changed to an enhancement mode of operation by the effects of the N^+-doped poly gate.

The reduction in overlap capacitance that was made possible with this *self-aligned* silicon (poly) gate process increased the speed of the logic circuits by a factor of 5:1. During this time period, the technologists were also learning more about the physics of the surface of silicon and there was now strong motivation to develop an NMOS process to take advantage of the 3:1 performance benefits of the N-channel transistor over the P-channel transistor.

2.3 NMOS FOR HIGHER SPEED

There were many reliability problems with the first NMOS products, but there were many benefits to the system designer that resulted from NMOS logic curcuits. For example, in circuit design the N-channel transistor is similar to the NPN bipolar transistor, and logic circuits built from either of these transistors can be operated from a single 5-volt power supply.

Using a Substrate Bias

Many of the early NMOS products made use of a negative voltage to bias the substrate rather than simply grounding this region. The use of substrate biasing has two benefits: (1) The magnitude of this bias voltage can be automatically regulated to control the threshold voltage of the N-channel transistors, because of the *body effect* or substrate effect on threshold voltage, which prevents an undesired shift to depletion mode. (2) Biasing the substrate also raises the breakdown voltage of the transistors. Some NMOS products still make use of on-chip-generated substrate biasing to obtain this higher breakdown-voltage advantage.

Logic Circuit Benefits

The conversion to NMOS meant that negative power supply voltages were no longer needed. NMOS logic circuits were designed to operate from the standard 5-volt T^2L power supply voltage. The NMOS logic voltage levels were easily designed to be directly compatible with T^2L logic voltage levels.

The on-chip power dissipation associated with a digital output buffer could also be significantly reduced when compared to a PMOS output buffer. This reduced chip dissipation allowed the designers of microprocessors to add additional circuitry to the chip, greatly extending the capabilities of the NMOS microprocessors.

Problems in Realizing Linear Circuits

The design of modern bipolar linear ICs has been greatly aided by having both NPN and PNP transistors on the same chip. Both of the single-channel MOS technologies, PMOS and NMOS, have increased the difficulty of linear circuit design, although special linear functions have been designed using an NMOS process where the op amps have had to meet only limited performance requirements. Neither of the single-chanel MOS technologies has provided general purpose linear circuits that could compete with the performance obtained with the standard bipolar linear products.

2.4 THE BIRTH OF CMOS

Early complementary MOS (CMOS) products were mainly used in systems where low power drain was important, although the dominant share of SSI and MSI logic was still handled by T^2L and low power Schottky T^2L products. CMOS logic therefore served in specialized applications but did not enjoy the large volume of the well-established bipolar logic products because CMOS products were slower and also more expensive.

Logic with No DC Power Drain

The unique thing about CMOS logic circuits is that the presence of the complementary transistor type allows the design of logic circuits that *do not consume dc power*. This can be seen in the logic inverter circuit that is shown in Figure 2-10. This power saving feature results because both of the transistors cannot be conducting current at the same time. (We neglect the small rise and fall times, where simultaneous conduction does occur.) For example, if the logic input is in the 0-volt state, the N-channel transistor will not be conducting and only the P-channel transistor will be ON. This ties the low resistance of the P-channel transistor directly from the power supply line to the output node of the circuit. All of the

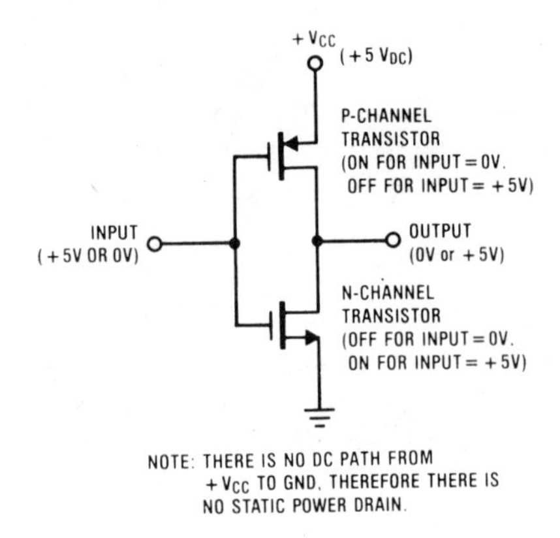

Fig. 2-10 A CMOS Logic Inverter

stray capacitance attached to this output node will be charged essentially to the power supply voltage level, but *there is no path for dc current flow* in this logic state.

If the logic input is, instead, at 5 volts, the P-channel transistor will have a 0-volt gate-source drive and will be OFF. The N-channel transistor will now be ON and this will tie the low ON resistance of this transistor from the output node to ground. All of the stray capacitance at the output of the logic circuit will be discharged to 0 volts and, again, *there is no path for dc current flow*. Notice that when the N-channel inverter transister is ON, the load device has switched OFF. This unique CMOS *gated-load* concept is the reason for the low power drain of CMOS logic.

Notice that if the input voltage were to be held at 2.5 volts, both transistors would simultaneously be above threshold and dc current would then flow from the power supply line directly through these two transistors in series to ground. This glitch of current flow during the small rise and fall times is typical for most logic circuits and is one of the components of power dissipation for a CMOS logic product. In addition, notice that the stray capacitances are first charged to the power supply voltage level and then discharged to ground during the next logic state.This *displacement current flow* into these capacitors also represents a switching-frequency-dependent component of power dissipation for a CMOS logic product. Both of these causes of power dissipation increase as the clocking frequency of the CMOS logic is increased. If the logic circuits are not changing state, the dc power drain will drop to essentially zero because only the junction leakage currents will flow.

The power drain of CMOS products can be calculated by making use of an equivalent power dissipation capacitance value, called C_{pd} and specified on the data sheets, that accounts for both the rise and fall time current glitches and the internal stray capacitances of the logic circuitry. Power drain (P_d) can be estimated by using the following relation

$$P_d = (C_{pd} + C_L)(V_{CC})^2(f_{clk}) + (V_{CC})(I_1)$$

where: V_{CC} = Power supply voltage
I_1 = Leakage current
C_L = External load capacitance at the output
f_{clk} = Clock frequency

Notice that the power drain varies as the square of the power supply voltage. Therefore reducing the power supply voltage from 5 to 2 volts, assuming the speed of the logic is still adequate, will reduce the power drain by $5^2/2^2$-25/4, or 6:1. If the clock frequency is forced to zero during standby intervals, the only source of power dissipation is the leakage current of the junctions. Many times the leakage current listed on the data sheet is larger than the actual current that exists to make it easier for the IC testers to measure.

The low power benefit of CMOS does not result if the logic circuits are rapidly switched. In this case, the power drain depends on the stray capacitance that must be driven and becomes nearly the same for NMOS, bipolar, or CMOS logic. There is generally a significant system power reduction benefit when CMOS logic is used, because only a relatively small percentage of the total logic circuits in a system will be toggling at the maximum clock rate.

The Early Metal Gate CMOS Process

The first CMOS process added N-channel transistors to the existing PMOS process. Separate areas of lightly-doped P-type regions had to be created on the surface of the silicon, as shown in Figure 2-11. Extra processing steps were added to fabricate the N-channel transistors. The first step was to form the *P-wells* by a carefully controlled light impurity-doping that would invert the N-type substrate, yet not provide too heavy a concentration of impurity dopant in these P-wells. Following the formation of these P-wells, the sources and drains for both transistor types were provided by using two, high doping concentration, diffusion steps.

The added processing steps that were required for a CMOS IC raised the cost of fabricating a CMOS wafer and this was reflected in higher costs for the finished products. The early CMOS processes used an aluminum gate and the low priority usually given to the CMOS processing lines limited the process updating. With the onset of the present VLSI era, CMOS processes are now receiving the most attention.

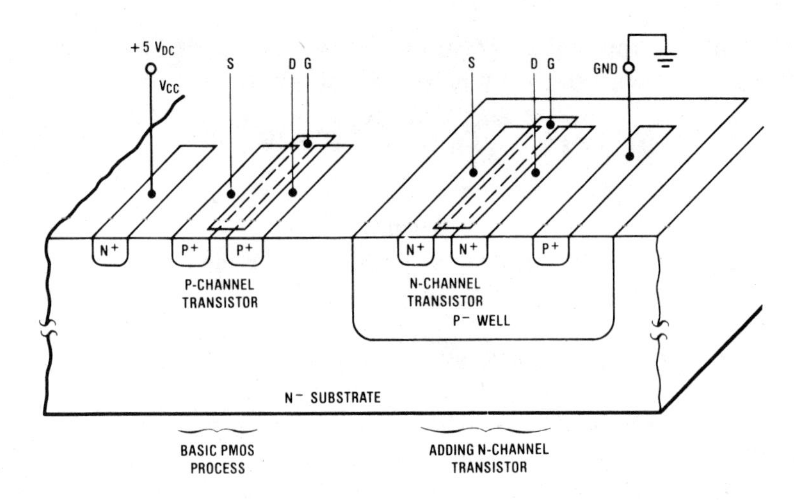

Fig. 2-11 Adding N-Channel Transistors to Make CMOS

Transmission Gate Logic

In IC design, the details of the logic circuits vary drastically as the IC technology changes. One of the first of these examples was the strange looking multiple-emitter input transistor that is characteristic of bipolar T^2L logic. A unique thing about CMOS logic circuitry is the widespread use of analog switches, which the digital designers call *transmission gates.* These analog switches, shown in Figure 2-12, have the ability to operate with input voltages that can even range slightly outside the power supply voltage range (-50 mV to $V_{CC} + 50$ mV). Limits exist at one diode forward-voltage drop below ground and one diode forward-voltage drop above the power supply voltage.

The CMOS transmission gate handles this wide range of input voltages by using two complementary transistors that are wired in parallel.

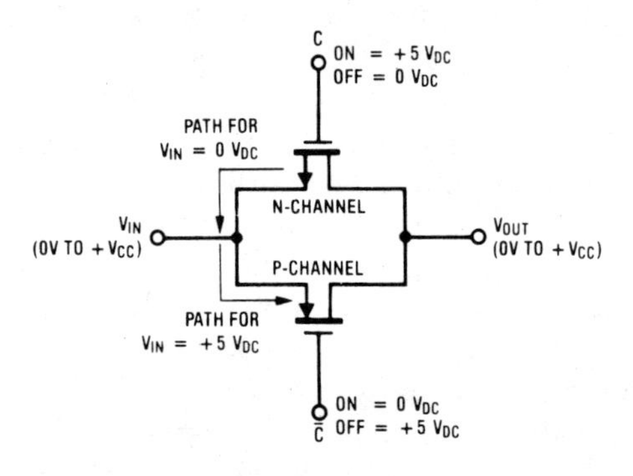

Fig. 2-12 The CMOS Transmission Gate or Analog Switch

Both of these transistors are controlled by clock inputs: clock (C) and the inverted clock ($\overline{C}$). Low input voltages are handled by the N-channel transistor and high voltages are handled by the P-channel transistor. Notice that when the *input voltage* to this transmission gate is at either 0 or 5V, one of the transistors has a *maximum* gate overdrive voltage (the other transistor is OFF) and so provides a low ON resistance switch closure. The largest value of ON resistance for a CMOS transmission gate occurs when the input voltage is one-half of the power supply voltage, approximately 2.5V for a 5-volt supply. Although both transistors are ON, neither has a large gate-source drive voltage.

This ability to operate over a wide range of input voltage is a major benefit of the CMOS transmission gate when compared to an analog switch that is built using a single channel technology, such as NMOS. With only the N-channel transistor available, the maximum input voltage must be restricted to 2 or 2.5 volts to allow a gate voltage of 5 volts to turn the transistor ON. If operation with a 5-volt input is required, a higher-voltage gate drive must be used. This extra supply voltage has to be externally provided to the IC by the user, or generated on-chip with a capacitive voltage booster.

As an example of the use of transmission gates in logic design, Figure 2-13 indicates a CMOS *flow-through latch*. The two switches shown in this diagram are operated out of phase: when one is open, the other is closed.

To put this latch in the flow-through mode, SW2 is opened and SW1 is closed. With SW2 open, the lower logic inverter is disconnected and the circuit is not allowed to regenerate. Any logic that is applied to the D input *flows through* the single inverter to provide the $\overline{Q}$ output signal.

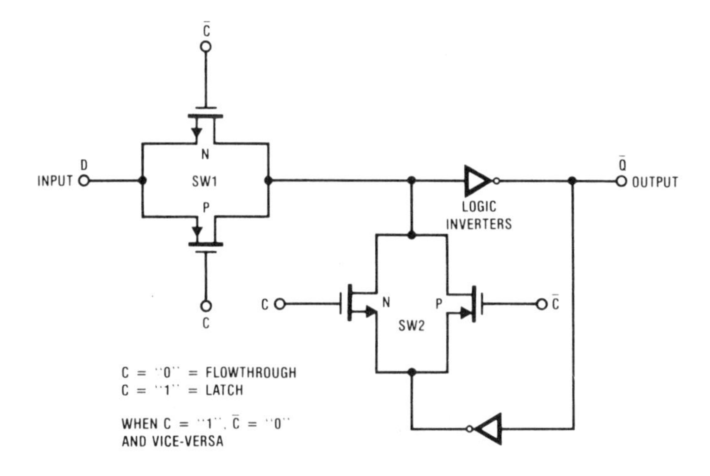

Fig. 2-13 A CMOS Flow-Through Latch

The present logic state can be latched by changing the logic level of the clock. The switch, SW1, at the input then disconnects the D input; and, when SW2 closes, the two inverters form a regenerative flip-flop circuit. *This is the latched state.*

Two of these flow-through latches, connected in cascade, provide the useful, edge-triggered, D-type flip-flop. This use of transmission gates to build static logic circuits is unique to CMOS products.

Static RAMs that Can Be Put to Sleep

When CMOS logic is not clocked, the power drain drops to the very small junction leakage level, and all of the internal logic states are maintained. This is made use of in CMOS static RAM products. This memory operates normally with the power supply voltage at 5 volts. The contents of the memory can be retained indefinitely by reducing the power supply voltage to approximately 2 volts. In this *memory retention mode*, only the junction leakage currents flow. Read and write cycles are not permitted while in the retention mode. This useful memory, that only draws leakage currents, is unique to CMOS.

Benefits and Problems Because of the Extra Devices

The added processing complexity that is used to fabricate CMOS products also provides parasitic complementary bipolar transistors. The parasitic NPN transistor, shown in Figure 2-14, has often been used as an emitter follower, because the collector is inherently connected to V_{CC}, to provide high output current for CMOS products. These NPN bipolar emitter followers are also useful in CMOS analog designs.

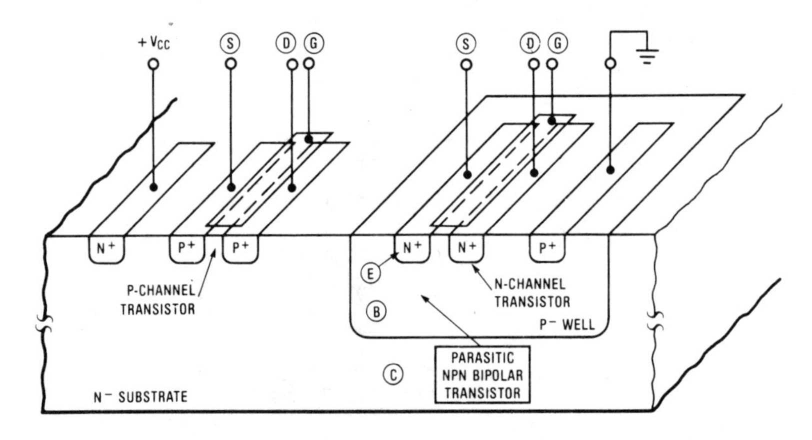

Fig. 2-14 The Parasitic NPN Bipolar Transistor in CMOS

Problems exist because this NPN bipolar transistor can couple to a parasitic lateral PNP bipolar transistor and form the undesired *hook connection* or *Silicon Controlled Rectifier* (SCR) that is shown in Figure 2-15. This is the reason for the *latch-up* problem with CMOS. This latch-up mechanism establishes a low resistance path from the power supply pin to ground because of an internal SCR action. Uncontrolled amounts of current can flow—limited mainly by the external power supply—which can melt the metallization on the IC chip, melt the bonding wires, exceed the power dissipation of the package, or, at minimum, hold the logic circuit in this inoperative, latched-up state until the power supply voltage is cycled, switched OFF and then back ON again.

To trigger this SCR, a lead of the IC must be taken to a negative voltage of one forward-diode voltage or be taken above the power supply voltage by one forward-diode voltage. In addition, a sufficient amount of current must be input to the IC or removed from the IC to achieve the necessary conditions for a regenerative latch-up within the parasitic bipolar transistor hook connection.

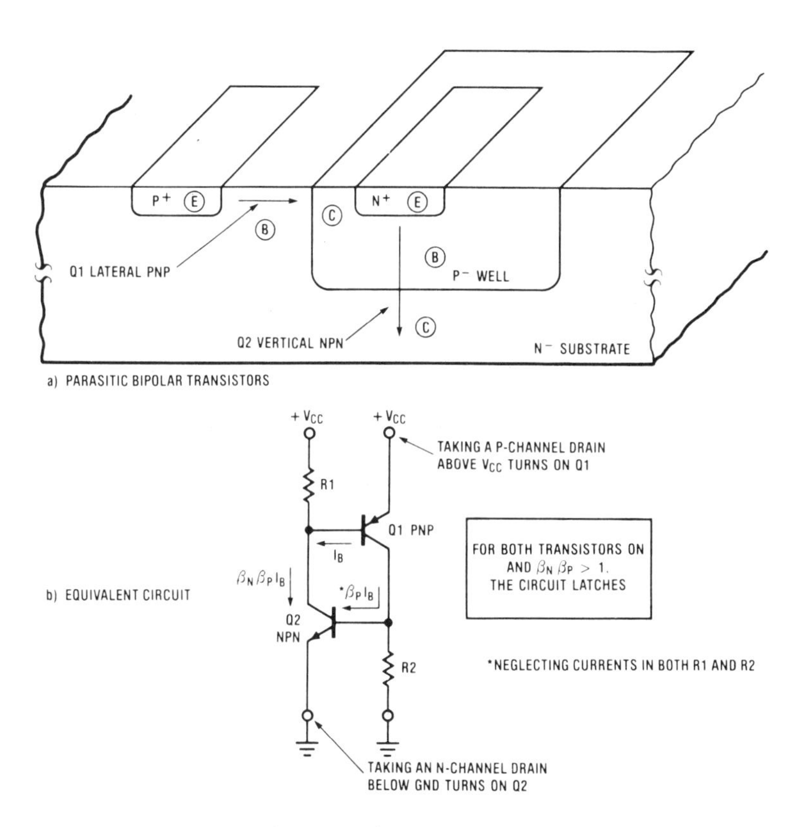

Fig. 2-15 The SCR in CMOS

The simplicity of the single channel MOS fabrication does not provide these parasitic complementary bipolar transistors and therefore this latch-up problem is unique to CMOS products. Parasitic SCRs are also possible within both linear and digital bipolar IC products.

Special chip layout techniques can reduce the current gains and/or *collect* the current of each of these parasitic bipolar transistors, but a more straightforward solution to this problem is the use of a richly-doped substrate (<1 Ω-cm) with only a relatively thin, lightly doped (>30 Ω-cm) epi layer on top for the fabrication of the MOS transistors. By making use of this epi-substrate, CMOS logic products have been built that guarantee no SCR problems for ± 20 mA of current forced into or out of any input or output pin. This guaranteed spec applies over the complete operating temperature range. The reduction in SCR susceptibility results because R1, of Figure 2-15b, is significantly reduced in value.

This epi-substrate SCR reduction technique works somewhat easier with P-well processes because these make use of an N^+ substrate, as shown in Figure 2-16. To prevent the rich doping of the N^+ substrate from diffusing up into and undesirably enriching the lightly-doped N^- epi layer, the slowly diffusing N-type dopants arsenic or antimony are used to form the N^+ substrate.

There are three N-type dopants that are used in IC fabrication: phosphorus, antimony, and arsenic. The last two of these are very useful because they have slower rates of diffusion in silicon. Processes can be designed where N-type regions will essentially stay put after they have been diffused into the silicon.

For P-type doping in IC fabrication, only boron, a fast diffusing dopant, is widely used. P^+ layers tend to be avoided in processes because they move too much during high temperature exposure in subsequent wafer processing. This adds somewhat to the difficulty of using epi-substrate SCR reduction in N-well CMOS processes. (N-wells versus P-wells will be discussed later in this chapter.)

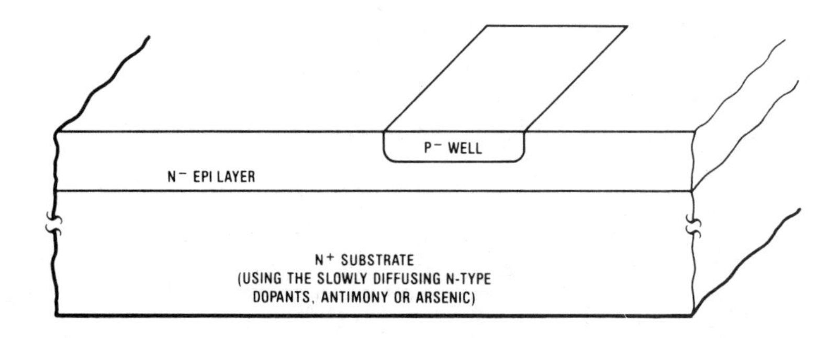

Fig. 2-16 The Epi-Substrate for Reducing the SCR Problem

SCRs are also sometimes triggered by *impact ionization* because of the resulting *substrate current flow*. In the depletion layers of the drains, mobile charge carriers that enter from the channel region of the MOS transistor are accelerated by the high value of the electric field strength that exists in this space charge layer. In logic circuits this high value of electric field only occurs during switching. These mobile charge carriers collide with the lattice atoms, thus creating hole-electron pairs and *photon emission* on impact with the lattice, as shown in Figure 2-17. This is similar to *avalanche breakdown*, but impact ionization takes place at a lower, nonregenerative level of current flow.

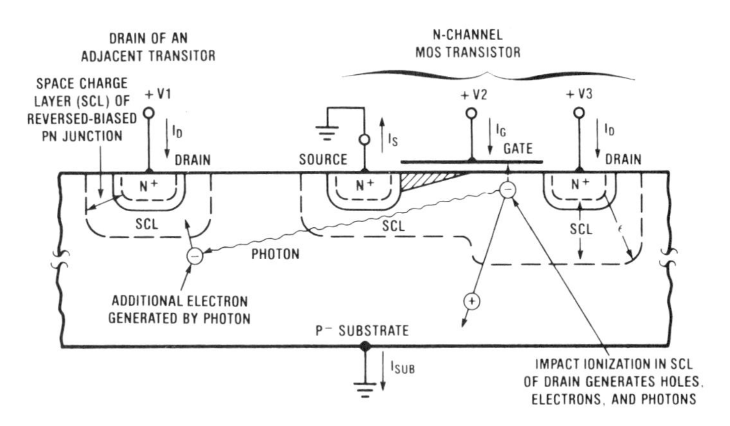

Fig. 2-17 Impact Ionization in a Short-Channel MOSFET

Impact-generated holes are swept into the substrate. This substrate current flow through the high bulk resistance within the substrate region creates substrate voltage disturbances that can potentially trigger an SCR, or at least degrade circuit performance.

These impact-generated electrons can also be *trapped* within the *gate oxide* and alter the surface states. Both of these possibilities will eventually raise the threshold voltage of the MOS transistor. Finally, remote *free electrons* can be provided by the impact-generated *photons* and these free electrons can be collected by, and therefore disturb, other MOS transistor circuitry that is in proximity on the IC chip.

The impact ionization coefficient for electrons is approximately three orders of magnitude larger than that for holes; therefore, this substrate current problem is much greater for N-channel than for P-channel transistors. This phenomenon has caused some researchers to favor P-type substrates (N-well CMOS) as gate lengths are scaled below 2 μm (where 1 μm $= 10^{-6}$ meters) because the substrate current that results from impact ionization will be more effectively shunted than in the case where only a P-well exists, as in P-well CMOS.

A Useful Zener Diode Exists in CMOS

A zener diode can also be provided in a CMOS product. This was first used on an 8-bit analog-to-digital converter that made use of a metal gate CMOS process.

To reduce the chip area, the heavily doped N^+ and P^+ diffusions that form the sources and drains of the complementary MOS transistors are allowed to touch each other. This is called *butted guardbands* and any of the older metal gate CMOS products that limit the maximum power supply voltage to approximately 6V typically use this die-area-saving layout trick. This P^+N^+ junction forms a zener diode that has a breakdown voltage of approximately 6.5V. To guarantee that only the desired zener will conduct, and not all of the other P^+N^+ junctions that exist in a butted guardboard layout, the circuit shown in Figure 2-18 has been used. Notice that the lateral PNP transistor has a collector-to-base short to form a diode connection that will subtract a forward-diode voltage drop from the power supply voltage that is supplied to the rest of the CMOS chip. This reduced power supply voltage prevents the possibility of zener breakdown within the logic circuitry. This has been a useful voltage regulator for many low cost systems.

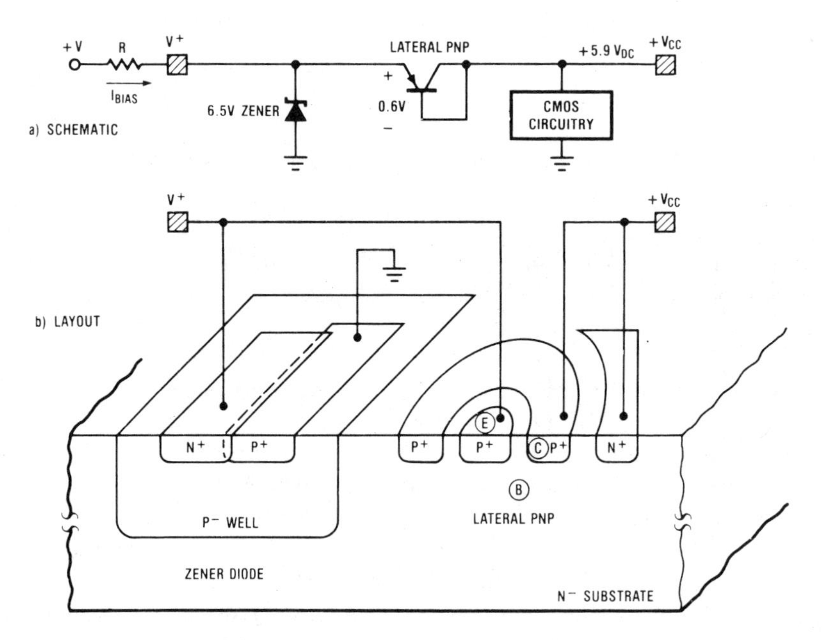

Fig. 2-18 Building a Shunt Voltage Regulator in CMOS

N-Wells Versus P-Wells in CMOS

The early CMOS processes were derived from PMOS processes and therefore used the same N-type starting substrate. The P-channel transistors were still formed directly in the surface of this substrate, as in PMOS fabrication, and a separate P-well was added to contain the N-channel MOS transistors. This created a richer background doping level for the N-channel transistors, which compromised their electrical performance. The carrier mobility is slightly reduced, the junction capacitance is increased, and the substrate biasing effect—the body effect on threshold voltage—is increased. Raising junction capacitances directly adds to the parasitic stray capacitances which increases the propagation delay. The increased body effect reduces the transistor performance whenever the source is operated at a voltage level above ground, such as is done with transmission gates.

Other CMOS processes for logic circuitry have interchanged the doping types and start with a P-substrate, the same as is used for NMOS products. These processes built on the existing knowledge of advanced NMOS processes and NMOS memory cells, and the switching speed is improved because the N-channel transistor is now favored. Now an N-well is used, so this is often called a *reverse-well* CMOS process.

The wafer processing can even be shifted from P-well to N-well, and vice versa, with no externally noticeable changes in the packaged IC logic products. Problems do occur if a circuit design happened to make use of the parasitic NPN bipolar transistor that is available in a P-well process because the parasitic bipolar transistor that is available changes to a PNP in an N-well process. The parasitic NPN transistor has higher β and higher current capability than the parasitic PNP transistor. This makes the N-well CMOS process have an inherently better SCR latch-up immunity. We will have more to say about the SCR problem in CMOS in Chapter 5.

Linear circuits often make use of these parasitic bipolar transistors and cannot easily be changed from P-well to N-well processes or vice versa. There is a benefit with P-well CMOS processes for linear circuits because the background, or body, for a particular N-channel transistor can be biased at any desired dc voltage. This removes the body effect on threshold voltage in linear designs and also allows these transistors to be cascaded in high voltage applications.

The selection of N-well versus P-well for a CMOS process involves many conflicting requirements. Dual-well CMOS processes optimize both transistors at the expense of a more complex process.

CMOS Is a Natural for Linear Circuits

Linear MOS IC design has been aided by the development of many unusual parasitic device structures within the CMOS processes. The single-channel MOS technologies do not have any interesting parasitic junction possibilities and there is also only one type of MOS transistor available for the IC designers (although both depletion and enhancement mode transistors may be available). This lack of extra devices in an NMOS process is a major problem for MOS linear designs. The extra junctions available in the CMOS processes make many additional parasitic structures available to the circuit designers. In fact, the recent interest of the linear designers in CMOS processes has increased the number of different diodes and transistors that have been tricked out of the basic CMOS process.

Linear IC design has come to depend on the availability of complementary bipolar transistors. Therefore, the complementary nature of CMOS has been attractive to linear IC designers. Many of the new linear circuits that are appearing in CMOS are basically using the interchange of an N-channel MOS transistor for a bipolar NPN and a P-channel MOS transistor for a PNP in the traditional circuits that have been previously developed for bipolar linear products. It is interesting to notice that the Darlington connection of two transistors, which has been useful to greatly increase the current gain when compared to a single bipolar transistor, is not needed in MOS designs.

In addition to the conventional linear circuits, the availability of excellent, low-cost, analog switches and high-quality MOS capacitors has created an interest in *sampled-data systems.* This is an interesting historical development because linear circuits got started with chopper-stabilized circuits in the bygone days of the vacuum tube. There are many similarities between vacuum tubes and MOSFETs. To overcome the poor offset voltage that also existed with vacuum-tube op amp circuits, mechanical choppers were used. These chopper-stabilized vacuum-tube op amps were the workhorses of the early analog computers.

Today, we are returning to these chopper-stabilized design techniques and have replaced the mechanical chopper with CMOS analog switches. We now call this technique a *sampled-data* approach. The analog switch, the transmission gate we discussed earlier, the parasitic bipolar transistor emitter follower, and the zener diode are all useful to the linear CMOS IC designers.

A current mirror (see Figure 1-14) can also be made using MOS devices. The match in threshold voltages between two MOS transistors on an IC chip allows one of these transistors to be diode-connected (i.e., a gate-to-drain short) and the resulting gate-source voltage of this transistor is applied to a second MOS device to control the current flow in the drain of this second transistor. Current scaling can also be accomplished, if desired, by scaling the W/L ratios of these two MOS transistors. This W/L scaling is similar to the emitter area scaling that is used in bipolar designs. Further, the base currents, which constitute an error in a bipolar current mirror and have forced the use of a Darlington connection for highest accuracy, are no longer a problem with MOS-FET transistors. Therefore a simple, two-transistor mirror is all that is needed.

A relatively recently discovered device also exists within a CMOS process: the parasitic lateral NPN bipolar transistor, shown in Figure

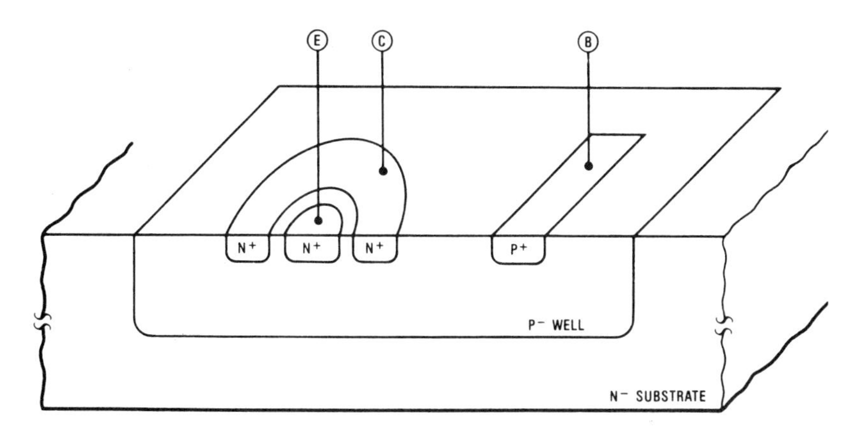

Fig. 2-19 The Parasitic Lateral NPN Bipolar Transistor

2-19. This device is useful to provide voltage references and for biasing circuits in modern CMOS linear products (as we will see in Chapter 5).

A general problem facing the use of linear products is that digital electronics continues to replace linear systems. This trend is continuing at an increasing pace. Departures from an analog solution are often made because changes in a digital system are easier to implement. For example, the slight shifts in the performance parameters and the options that are needed to operate with a wider range of end-use systems can be handled by a single digital IC chip where the customization and changes are made by program alterations in an on-chip ROM. In addition, logic circuits can be more easily handled by computers in both design and automatic layout. (More on this in Chapter 4.)

A further problem for linear circuits is the continuing reduction in the magnitudes of the system power supply voltages. This loss of the capability for large signal-voltage swings reduces the signal-to-noise ratio of low-voltage linear systems and also restricts the dynamic range. It may seem that the threshold voltages of the CMOS transistors could be reduced to allow somewhat larger analog voltage swings. Unfortunately, a lower limit exists at approximately 0.5V on threshold voltages to reduce the problem of subthreshold current flow in the MOS transistors.

There are also simplifications in testing digital circuits as compared to testing linear circuits. It is still expected that the optimum system solution will be a mixed solution, which takes advantage of both analog and digital approaches. CMOS processes and bipolar with CMOS processes have much to offer for combining this circuitry on a single chip.

Bulk CMOS and Silicon on Insulators

Most of the CMOS effort in the industry has involved CMOS that is fabricated on the surface of silicon wafers. This *bulk CMOS* has been continually challenged by researchers who favored growing an epitaxial film of silicon on the compatible crystal structure of an insulating sapphire substrate called *CMOS SOS*, Silicon on Sapphire.

Many speed-power advantages are realized with these CMOS SOS products, but the major semiconductor manufacturers have never adopted this technology. Recently, even some of the earlier supporters of SOS are moving back to bulk for their advanced CMOS products, because speed is more limited by interconnect wiring capacitance. Although many observers have continually predicted the death of SOS or the more generalized *Silicon On Insulator* (SOI), technologies continue to be reported that make use of this concept, especially for military and aerospace applications. It is expected that the advanced CMOS processes may, at last, stop the interest in SOI for commercial applications.

2.5 THE NMOS REVOLUTION

When the early reliability problems of NMOS were solved, this technology made rapid improvements in logic circuit performance. It appeared that this was the ultimate in MOS technology and a wide range of high performance NMOS products was quickly produced.

The Greed for Speed

With the switch to the high performance N-channel transistor and the performance benefits of a self-aligned gate process, the technologists pushed in other directions to find ways to further improve the speed of MOS logic circuits. The goal was to catch up with, and even try to exceed, the performance of bipolar logic.

Some early researchers, rather than requiring expensive high-resolution lithography, built on the basic trick of providing a narrow basewidth in bipolar transistors by making use of differences in diffusion depths. This resulted in both *Diffused MOS* (DMOS) and a novel *Vertical MOS* (VMOS) structure.

The idea of DMOS, Figure 2-20, was to make use of a single oxide opening, or window, through which two separate diffusions of opposite doping types were introduced. A short channel length for this DMOS transistor resulted because of the differences in the lateral diffusions of these regions.

The VMOS device, Figure 2-21, makes use of a thin epi layer to form a narrow channel length in the vertical dimension. In addition, a special etch is used that follows the crystal structure of the wafer and creates the "V" depressions in the surface. This provides a wide channel width: the perimeter of the resulting four-sided structure.

While these tricks were being pursued, other researchers directly attacked the problem of providing higher resolution in the lithography

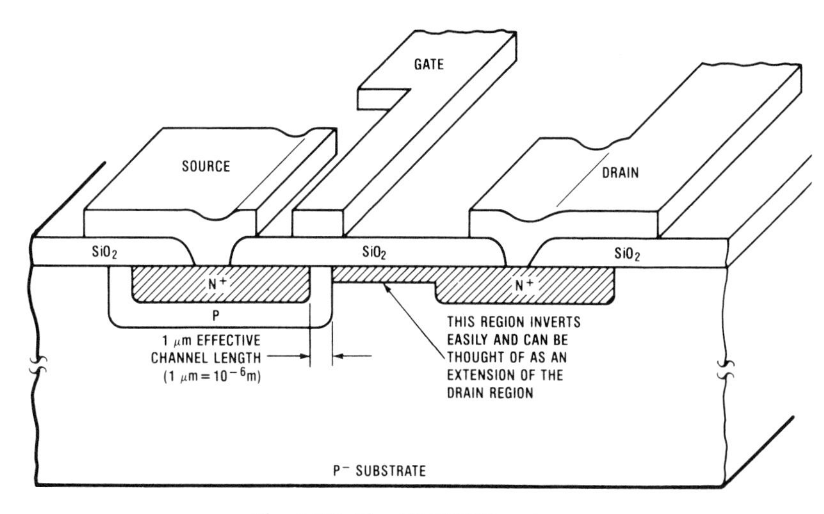

Fig. 2-20 The DMOS Structure

used for IC fabrication. The success of this latter group has kept DMOS and VMOS out of general IC production, but these techniques have been used to create a new market—discrete power DMOS transistors.

There is finally strong interest in taking advantage of the successful discrete DMOS power transistor in single-chip, linear CMOS power ICs. It is interesting that when the structure of the discrete power DMOS transistor was modified from that shown in Figure 2-20 by making use of

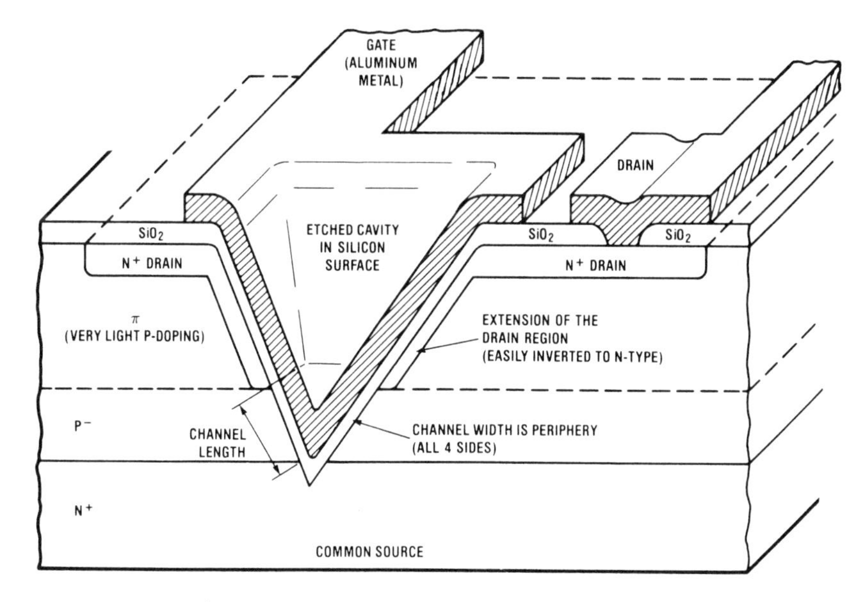

Fig. 2-21 Cross Section of a VMOS Transistor

only sources and gates on the surface, with the drain at the bottom, unusually high voltages (for a given channel length) could be handled. The increased vertical region that is available for the reversed-biased depletion region spreads in this vertical DMOS structure provided this advantage over conventional CMOS transistor structures that are built on the surface of the chip. The further advantage of the more rapid ON and OFF switching times that can be achieved with a MOS "unipolar" transistor, as compared to a "bipolar" transistor (because of the absence of minority charge carriers) has become a very desirable characteristic for the widespread customer interest in high-efficiency ON-OFF or duty cycle modulated power circuitry.The DMOS transistor therefore provides a power transistor that can be rapidly switched ON and OFF, handles very high voltages relatively easily, naturally decreases current flow at high chip temperatures (which prevents hot spot formation), and has no safe operating area limitations. These almost ideal power transistors are now being merged with the low-level control circuitry to make a new family of linear power ICs. These products do not require the smallest feature sizes; but if small features are used, the surface density of the source regions can be increased, which reduces the ON resistance of the transistor and therefore raises the efficiency: the transistor becomes a more ideal switch. This means that larger chip sizes with smaller feature sizes result in less power dissipation on the chip, which improves system efficiency and also allows lower-cost packages with poorer heat sinking capability to be used. For "bang-bang" servo applications, these power DMOS IC products will be strong competitors.

Benefits of Device Scaling for Digital Circuits

Improvements in lithography allowed device scaling in digital circuits where all of the horizontal and vertical dimensions associated with the MOS transistor were reduced by the same scaling constant, K. Actually, deviations in this exact scaling do occur depending on the IC application, the desired performance, and the technological capability. If exact scaling is used, the electric fields within the device and the power dissipation per unit area of the device will remain constant as the feature sizes are scaled to smaller dimensions. Other characteristics vary with the scaling constant, K, as:

Characteristic	Scales As:
Doping concentration	K
Voltages (V) and Currents (I)	1/K
Capacitances (C)	1/K

This results in:

Propagation Delay	$RC = \frac{VC}{I}$	$1/K$
Power (VI)		$1/K^2$

and therefore,

Power-Delay product	$1/K^3$

So if a value of $K = 3$ can be achieved, it will reduce all sizes by a factor of 3, and the power-delay product will drop by a factor of 9:1! This assumes that all of the scaling is done, even reducing the power supply voltage from the standard to 5V/K.

The production of scaled VLSI devices requires control and minimization of detrimental short-channel effects. These are the reduction in both the threshold voltage and the carrier mobility in the channel and therefore g_m reduction. The reduction in threshold voltage exists with short-channel transistors because of coupling of the electric field that is established by the potential applied to the drain. This *second gate* effect is generally not a problem for channel lengths down to 2 μm; but below this, special processing steps must be added to minimize this threshold voltage reduction problem.

Measurements can be made on an MOS transistor to determine if short-channel effects are taking place. For example, in a long-channel device, the drain current is inversely proportional to the channel length. If a device geometry conducts more drain current than is indicated by this relationship, it is exhibiting short-channel behavior. Also, in the subthreshold region of operation, a long-channel device will provide an essentially constant value of drain current as V_{DS} is varied for $V_{DS} > 3\frac{kT}{q}$. Short-channel devices show a drain voltage dependence on I_D for all values of V_{DS}.

To minimize these short-channel effects, a double implant is often used in the channel region. A deep, slightly higher dopant concentration implant is used to increase the substrate doping level in the channel region. This allows operating with higher supply voltages because it helps prevent the drain region depletion spread from reaching through to the source. A second shallow implant of the opposite impurity type is then used to adjust the final threshold voltage. The light original background doping of the body helps reduce the parasitic junction capacitance that exists because of the bottom and sidewall areas of the source and drain regions.

Advances in Wafer Processing

There have been continuous efforts to improve all aspects of silicon wafer preparation and processing. Contributions have been made by wafer suppliers, equipment suppliers, chemical suppliers, process researchers, and process engineers. The results of the combined outputs of these people have allowed the almost unbelievable achievements of the present VLSI era.

3.1 ION IMPLANTERS REPLACE DIFFUSION FURNACES

In most cases, modern silicon wafer processing makes extensive use of ion implanters. Ion implantation has replaced most of the older diffusion processes to introduce impurity dopants into the silicon crystal lattice. In the implant operation, *charged dopant atoms* (ions) are accelerated by an electric field and thereby obtain sufficient energy to penetrate the crystal lattice on impact. Accelerating voltages of up to 200,000 volts are used and result in dopant velocities in the range of hundreds of thousands to up to 4 million miles per hour.

This ion beam is designed to scan the wafer surface. The amount of doping can be easily monitored and controlled by simply measuring the current flow of the beam. Ion implantation takes place within a high-vacuum chamber and provides uniform and accurate doping concentrations. The implantation process can even penetrate thin-film barriers, such as SiO_2, that are used to prevent contamination at the edges of the MOS transistor gates, for example. Further, the doping profiles can be easily controlled by adjusting the ion energy, which affects the penetration depth. In addition, multiple implants can be done at temperatures that are low enough so that the existing doped regions within the silicon are not disturbed. After a number of implants have been made, one final relatively high-temperature cycle can be used to simultaneously activate and anneal the surfaces of all the implants. All of these benefits become important in the formation of the new shallow junctions that need very accurate doping control.

The implanted ions collide with the electrons of the crystal lattice, *electronic stopping*, or with nuclei of the target silicon wafer, *nuclear stopping*. Implant ions also can easily glide down the channels that exist within the crystal lattice under certain orientations of the wafer to the direction of the implant beam. If you have ever seen the typical three-dimensional physical models of crystal lattices that use balls to represent the atoms and sticks between these balls to represent the interatomic bonding, these clear channels exist at certain "viewing angles" where you can obtain an "unobstructed view" clear through the crystal model. If the ion implanting beam is allowed to "see" these unobstructed channels, the chances for collision are reduced and poor dopant control results because of the deep penetration of the dopant ions into the silicon wafer.

A collision between an incoming ion and a target atom results in a displacement of the target atom from its ideal lattice location. The target atom is knocked out of position. With high implant energy, this *bombardment* converts the top few thousand angstroms (e.g., 5000 Å = 1/2 micrometer (μm) = 1/2 micron) of the silicon surface into an amorphous layer. A high-temperature annealing operation activates the dopant atoms, allows them to substitute for silicon atoms in the lattice, and also nearly restores the overall crystalline order of the silicon, although many residual defects remain following the annealing cycle.

Large values of beam current for an implanter are in the range of 1 to 10 mA. Implant doses of 10^{16} atoms/cm^2 require more than 10mA of beam current. High current beams heat the wafer and create cooling problems. The early implanting machines were used mainly to lightly dope the channel regions of MOS devices and therefore required beam currents of only 0.2 mA. The move to complete use of ion implantation has pushed the maximum beam current closer to 20 mA and the energies have increased from 20 to 200 keV, although low currents are generally associated with the high energy implants.

The modern small feature sizes are requiring a change in the beam scanning method. The beam is no longer swept across the wafer surface; it remains stationary and normal to the wafer surface and the wafer is moved under it to eliminate *shadowing* problems (this is also done for very-high-current implanters because the beam cannot be steered without losing focus). The surprising thing is that the shadowing that is caused by the thickness of the resist layer can become significant in VLSI processing. For example, if the path length of the ions is 3 feet, a resist layer that is 1 μm thick can create a 0.08-μm shadow at the edge of a 6-inch diameter wafer. This is an example of the extremely complex equipment design problems that had to be solved to enter the VLSI era.

A new variation of ion implantation is to work with a cluster of atoms rather than a single atom and to then use this modified ion implanter to deposit films. This "Ionized Cluster Beam" (ICB) deposition technique puts a charge on, for example, up to several thousand atoms in one group. This cluster is then accelerated by a high voltage, but the energy of the beam (that is supplied by the acceleration of the ionized clusters) is spread over the atoms within the individual clusters.

High-quality thin films and interfaces can be formed at a relatively low substrate temperature with ICB. Also, if a high vacuum is used, single-crystal Al can be grown on a Si surface. This interface poses unusual structural and electrical stability against heat treatment and also has extremely high electromigration resistance. ICB is a deposition technique that may replace CVD, has good step coverage (as needed for the interconnect metal), is still experimental, and, unfortunately, has rather slow deposition rates.

This technique is very different from the use of 0.1- to 1-μm-size molten clusters, where instead of breaking up into individual atoms, the molten clusters splatter, like a liquid drop, on impact with the wafer surface. There is strong interest in developing this technology to allow the metal interconnect pattern to be directly written (no masks are needed) on each die to provide extremely fast turn-around time in gate array products.

3.2 BURIED DIELECTRIC LAYERS REDUCE STRAY CAPACITANCE

Researchers have formed buried dielectric layers to reduce the junction capacitance of the source and drain regions. Both oxygen and nitrogen ions have been used, although oxygen is favored because SiO_2 has only 60% of the dielectric constant of silicon nitride (Si_3N_4). Therefore SiO_2 dielectric layers provide smaller stray capacitances. Doses are 10^{18} ions/cm^2 and employ energies of 150 to 180 keV. To reduce the time required for this implant, approximately 100 mA of beam current has been used, which creates severe wafer heating and crystal damage problems. This is an interesting adaptation of the implant idea because now the implanted atom, O has been found better than O_2, *combines chemically* with the silicon *rather than replacing* the silicon atoms in the crystal lattice, as happens with impurity doping. Subsurface dielectric layers are highly desirable because circuit operating speeds are doubled, but many problems have to be solved before this can be put into large-scale production.

3.3 RESISTS

Resists are the link between the masking operation (lithography) and the etching operation. We will therefore introduce resists prior to the discussion of these other wafer fabrication steps. Modern lithography and etching are pushing the limits of resist technology.

Resists are polymer films that desirably have high sensitivity to an exposing radiation. Other characteristics of a resist are also important: good adhesion to the surface to be protected; high resistance to the effects of the etchant, the harsh chemicals and the mechanical bombardment of the new dry etches; good resolution capability, fine grain size; and good dimensional stability, no tendencies to creep.

The early resists were called "negative" resists and were polymerized by the action of incident ultraviolet light that passed through the clear areas of the IC mask plate. Following a masked exposure to light, the unexposed resist was dissolved away by a solvent rinse, the "developing" step. The resist film that remained on the wafer surface was then toughened by a bake cycle because temperature also would cause the resist to polymerize.

These negative resists would absorb liquids during the wet-chemical developing and etching steps. The resulting swelling of the film would distort the pattern. To combat this problem, and to reduce the grain size of the polymer, "positive" resists were introduced to allow patterning smaller feature sizes. This resist polymer is broken down by the incident ultraviolet light, opposite to the reaction of the negative resists. Working with these resists was different because this resist polymer could be *destroyed* if it was inadvertantly exposed to light prior to the completion of the etching step. The unexposed regions do not swell much in the developer solution, and so higher resolution is possible with positive resists. Therefore positive resists have replaced negative resists in most applications.

The ability of a resist film to achieve good linewidth control, high resolution, and good step coverage are mutually exclusive. Good step coverage requires a *thick resist layer*; high resolution and good linewidth control require a flat surface and a *thin resist layer*.

Researchers have proposed double and even triple level, tri-level, resist coatings. First, a thick organic layer (ordinary photoresist has been used) is placed on the silicon surface. An intermediate SiO_2 layer is then used. Finally, a thin top layer of x-ray resist is used, as shown in Figure 3-1.

This top layer is first exposed and developed. A plasma etch is then used to pattern the intermediate SiO_2 layer. Finally, the thick first photoresist

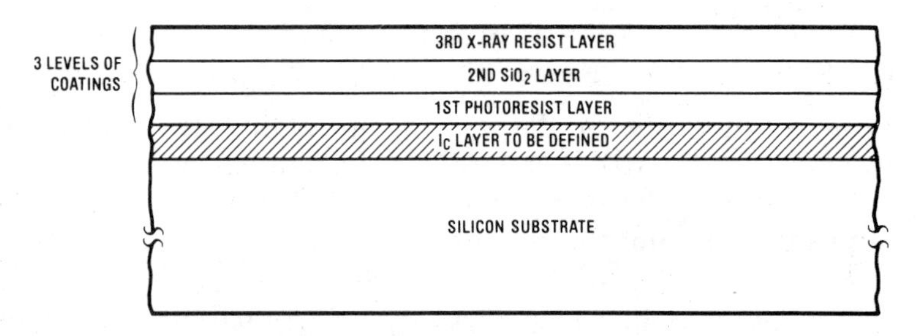

Fig. 3-1 Three Layers of Coatings Used in One Etching Step

layer is also plasma etched using the SiO_2 layer as a mask. Essentially vertical walls result and the submicron resolution of the x-ray system can be obtained at the wafer surface. This tri-layer photoresist technique can be adapted to any of the exposure systems. A general problem with these multi-level resist techniques is the high costs because of the added processing steps.

3.4 THE MOVE TO DRY ETCHING TECHNIQUES

In the past few years, the IC industry has been replacing the older wet-chemical etches with dry etching techniques. Dry etching is generally needed to define feature sizes less than approximately 3.5 μm. The major problem with wet etches is their inherent isotropic etching action, as shown in Figure 3-2. This causes undercutting. The material to be etched is attacked horizontally as well as vertically and is therefore also removed underneath the edges of the defining photoresist film. An additional problem is that the water-based etching solution can break the resist-to-substrate bonding. This *hydraulic wedging* action causes the edges of the photoresist film to lift off the wafer surface. Special primers are used as a very thin first coating to reduce this problem.

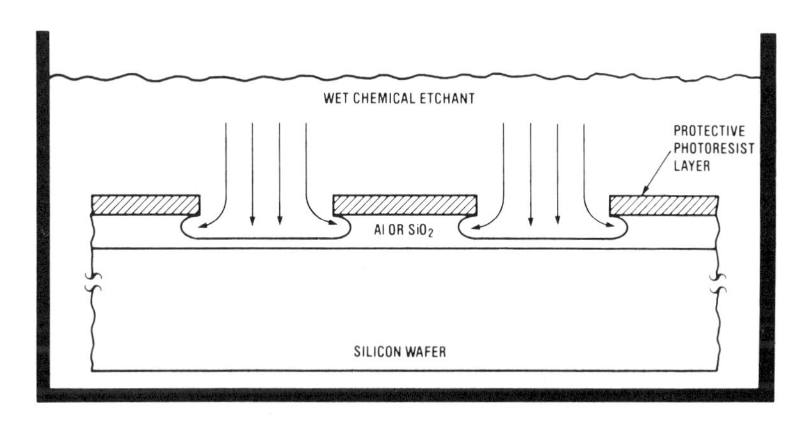

Fig. 3-2 The Isotropic Etching Action of Wet Chemical Etches

In VLSI processing, the ratio of the height (thickness) to the width of the etched film window (the *aspect ratio*) approaches unity, so the etchant can't be allowed to penetrate under the edges of the protective resist film. This has forced interest in anisotropic or dry etching techniques to form vertical etch profiles.

The benefit of dry etching is that a stream of high-energy charged ions is directed normal to the wafer surface to enhance the vertical component of the etching, as shown in Figure 3-3.

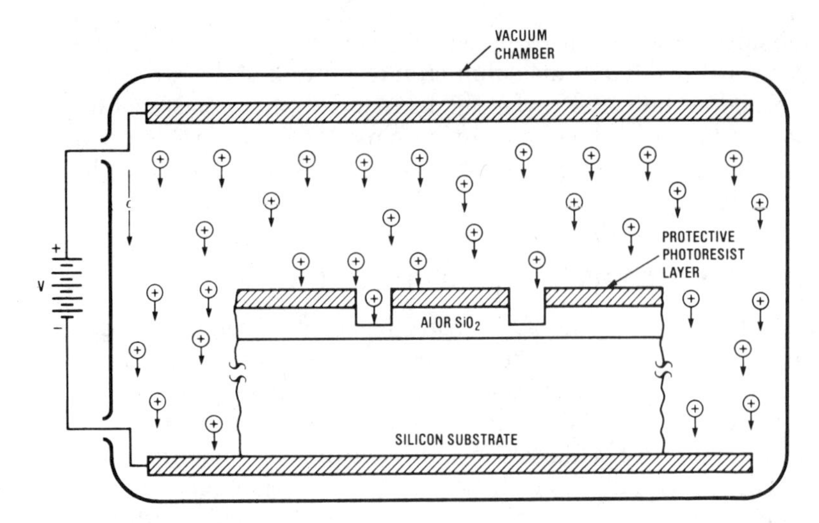

Fig. 3-3 The Nonisotropic Etching Action

Dry etching has produced 1 μm patterns with less than 0.05 μm of undercutting. A further benefit of dry etching is that the etching action can be abruptly stopped, unlike the continued etching of the wet processes that persists during wafer withdrawal and rinsing.

The dry etching techniques of today can be traced to the introduction of autoclaves (sterilization equipment) to the IC fab lines toward the end of the 1950's. These were initially used to remove the photoresist film following the etching step. Autoclaves employed high temperatures and high pressures in an enclosed quartz chamber.

To reduce the required temperature of the standard autoclave, a low vacuum (0.4 Torr) and a glow discharge was used. [In vacuum work, pressure is used to indicate the level or the degree of the vacuum and is measured in Torr. One Torr is 1/760th of atmospheric pressure, or atmospheric pressure is 760 Torr (14.7 lbs. per square inch)]. A *rough vacuum* is a pressure range of 760 Torr to 1 Torr. *Low vacuum* is a pressure range of 1 Torr to 10^{-3} Torr. *High vacuum* is 10^{-3} Torr to 10^{-6} Torr. (The vacuum industry also uses the micron as a pressure unit and 1 micron = 10^{-3} Torr. *Ultrahigh vacuum* is 10^{-9} Torr and below.)

A limited amount of oxygen was constantly pumped through this low pressure chamber. A glow discharge was established within the chamber by an rf field that was capacitively coupled from external metal plates through the walls of the quartz chamber. Capacitive coupling can be used because a rectification results because of mobility differences between the high speed, essentially massless electrons and the slower moving, more massive, positive oxygen ions. For example, during the positive cycle of the rf *sine wave*, the plate inside the chamber collects a large number of electrons. During the following negative cycle, a smaller number of positive ions is collected. On the average, a large number of electrons remains on this internal plate and thus, it becomes a cathode and sustains the plasma.

This rf induced glow discharge, or *plasma*, ionized the oxygen and created the very reactive nascent (recently formed) oxygen that more easily decomposed the organic films. The chemical results of this reaction were essentially all gases and could therefore be swept through the chamber by the constant gas flow.

This equipment was called an "asher" because it would volatilize organics and leave inorganics as an easily removed surface ash, similar to a self-cleaning home oven.

A disadvantage of the asher was that the relatively high concentration of metal atoms that were present in the early photoresist resins would be deposited on the surface of the wafer. This was a major problem for the early MOS ICs. A wet-chemical etch was therefore used following the ashing step to clean up these inorganic residues.

A big step forward came when gases, other than oxygen, were mixed and introduced into the asher to obtain improved chemical reactions. This resulted in the "barrel asher." The idea of chemically reactive gases that could etch films such as silicon nitride was developed, the *chemical ion etches*. It then became apparent that an etch should have a high degree of *selectivity*, that is, it should more rapidly etch the intended film but less rapidly etch the photoresist and the underlying layer. It was desirable to take advantage of this etch selectivity by making use of an intentional overetch. This would allow clearing the patterns on a large group of wafers in a single batch even though slight variations in etch time normally exist between the wafers. Therefore, a need for "end point detection" was created: a way to determine when the etch was essentially completed so a known amount of overetching could be used. By specifically looking for the chemical by-products of the etch, a fall-off on this chemical concentration (or a build-up in the amount of the underlying material) in the exit-gas stream was used to signal the end point and automatically stop the etching.

The next major advance was the parallel plate etcher. This machine established a plasma between electrode plates. These electrode plates were placed inside a horizontal reaction chamber where the wafers could be more optimally located for a more uniform etch. They were laid directly on one of the electrode plates. This horizontal wafer location did not allow as many wafers to be processed in a batch as compared to the older racks that held the wafers in a vertical orientation. These physical modifications improved the vertical nature of the etching. An rf excitation was still used because the frequency of this excitation was a useful processing parameter that was used to enhance the generation of particular etching species.

If the wafer plate was made the cathode, and the pressure in the chamber was reduced, a modified sputter etching would take place. Sputter etching is a *physical ion etch* where positve ions strike the wafer surfaces and etching results from momentum exchange. This modified sputter etching also had a large part of the etching action taking place as a result of chemical action. This is now known as Reactive Ion Etching (RIE).

In the modern etching equipment, a microprocessor controls all of the etching conditions, including the entrance of chemicals into the chamber. A programmed sequence can perform vertical etches or sloped etches on sequential IC layers during one continuous operation.

The *ion milling* machine is an interesting application of these etching principles where a directed narrow ion beam, instead of a large area plasma, is used in a medium-high vacuum system. Ion milling is useful in development labs and is used to etch difficult materials.

One of the biggest problems in fabricating ICs is the etching of the interconnect aluminum metal. This is not pure aluminum, but is rather a complex alloy that contains silicon (to prevent the aluminum from dissolving silicon from the IC chip and spiking or shorting through shallow junctions) and copper (to reduce the metal migration problem that exists with pure aluminum). This aluminum alloy favors the formation of an undesirably large grain structure in the aluminum film and also easily forms hillocks (random piles or small hills of aluminum). Both of these film defects are undesired for IC processing. The hillocks can be many times taller than the thickness of the aluminum interconnect metal film and will cause interlayer electrical short circuits. Large grain boundaries, that may extend completely across the width of the metal conductor, can create reliability problems as a result of metal migration (which will be discussed in Section 3.7).

To offset these problems, additional elements are added to the interconnect metal which creates a rather complex alloy. The reactive by-products of this metal etch should all be vapors so they do not interfere with the etching action and so that they can be easily removed from the vacuum chamber in the gas flow. The rather wide diversity of metals involved in this aluminum alloy make the selection of the etching chemicals rather difficult. In addition, these metal-etch chemicals tend to be very dangerous and difficult to safely handle.

In general, there is a move away from the older *batch fabrication* toward *single wafer fabrication*. A goal for all of the equipment is to process approximately 60 wafers per hour. There is also a trend toward fully automated wafer fabrication, including robotics.

3.5 THIN FILM DEPOSITION

In the processing of IC wafers, many types of both conductive and dielectric thin films are used. In the multiple layers of thin films that are used in modern processing, the stresses that are created within these films can cause cracking or crazing problems. The reduction of these stresses is a major part of the process development.

There is a close relationship between film deposition and film removal (etching). In some processes, an initial etch is often employed while the wafers are in the film deposition chamber to clean the surface just prior to the film deposition.

The early deposition of aluminum was accomplished by vacuum evaporators. Wafers were placed in a bell jar, a vacuum was established, and then filaments, that were previously loaded with short lengths of aluminum wire segments, were powered to melt the aluminum. The resulting aluminum vapor coated everything that was exposed, including a number of wafers that were placed within the bell jar.

The problem of the contamination that originated from the metals that were used for the filaments was solved by the move to electron-beam heating of the aluminum sample that provided the material that was evaporated. These "e-guns" used electric and magnetic fields to guide a beam of high energy electrons to the aluminum sample.

A move to sputtering systems allowed the deposition of complex materials that would disassociate in an evaporative deposition system. In sputtering, positive ions strike the sample and physically disrupt the material. It is this sputtering deposition equipment that easily allows the electrical polarities to be reversed to accomplish a slight amount of sputter etching of the wafers just prior to the sputter deposition of a film.

One of the most serious problems confronting VLSI is the reduction of particulate contamination. "Sputter-up" and "sputter-sideways" systems significantly reduce particles because they are drawn away from the wafers by gravity. Unfortunately, particles that originate from films on the internal surfaces of the vacuum system that happen to fracture during the deposition cycle may be electrically charged; therefore gravitational forces are easily overcome, and these "gravitational cleaning" techniques reduce, but do not eliminate, the particulate problem.

In the late 1950s, the first Chemical Vapor Deposition, CVD, processes were introduced to the IC fab lines. These depositions took place at atmospheric pressure in a reaction chamber where an organic gas would undergo simple pyrolysis, decomposition as a result of high temperature. A desired inorganic, that was a part of the complex organic gas molecules, would plate out on the heated wafer surfaces. The gaseous hydrocarbon residues of this reaction would exit in the gas flow. Both aluminum and dielectric films were produced by these early CVD processes.

Temperatures of approximately 700°C limited the usefulness of these processes because nearly completed IC wafers can't be exposed to such high temperatures. To reduce the reaction temperature, the pressure in the chamber was reduced to the 0.1 – 10 Torr range because the deposition rate is enhanced at reduced pressures. An additional benefit was that undesired small contaminating particles, which are 'airborne' at atmospheric pressure, simply do not exist at reduced pressures. The temperature of the wafers could now be reduced to the 500°C to 600°C range. Polysilicon and silicon nitride films were deposited by this Low Pressure Chemical Vapor Deposition (LPCVD) process because the reduced reaction temperature allowed these films to be applied to the wafers late in the wafer processing sequence.

To reduce the temperature still further, a plasma was excited within the reaction chamber. This is called "plasma-enhanced LPCVD." Temperatures of approximately 460°C could now be used, which is compatible with nearly finished wafers. This deposition technique is used for epitaxial silicon and polycrystalline silicon and also for the final wafer coatings such as silicon nitride (sometimes denoted as Si_xN_y to indicate that the exact compound is unknown) or, in general, oxynitride (to imply combinations of Si, SiO_2, and Si_xN_y) protective films.

The major difference between CVD and sputtering as a deposition technique is in the chemistry that is involved. Sputtering creates a supersaturated vapor which condenses on a cold surface. CVD uses volatile compounds as transport agents that contain the chemicals needed for the desired film. These volatile compounds react on a hot surface to plate out the desired film and the no longer needed chemicals are removed in the gas phase from the reaction chamber. CVD generally provides excellent step coverage and can also produce deposited films that will be selectively deposited on only metallic or "reactive" surfaces (such as silicon) and will not be deposited on insulator surfaces (such as oxides and nitrides). This selective deposition is self-aligning (and therefore improves process reliability) and simplifies the process because a masking step is not needed. CVD deposition of metals (especially tungsten) is an area of much current research and discovery.

Film deposition equipment is very sophisticated today and microprocessor control allows multi-layered film depositions to sequentially take place in a single processing step.

The best vacuum that is economically feasible is desired in order to provide deposited films with desirable and repeatable characteristics. Impurity atoms, which are also present within a vacuum chamber, continually strike the surface of the freshly deposited film and can react rapidly with the atoms of this film. Even if a chemical reaction does not take place, the presence of impurity atoms may affect the nucleation and growth of the desired film. This can increase contact resistance, reduce adhesion of the film to the chip surface, and also increase resistivity and grain size of the deposited film. In most cases high-rate deposition systems with the lowest impurity gas pressure will produce the highest-quality films.

An exciting technique, called Molecular-Beam Epitaxy (MBE), has been developed that allows the deposition of metals where an atomically abrupt interface exists. This deposition technique requires an ultra-high vacuum ($<10^{-9}$ Torr) and can be used to grow epitaxial metal layers on semiconductor surfaces. Good stable metal-semiconductor contacts result that are of interest in VLSI fabrication, and many advanced VLSI structures and many novel devices have also been fabricated, such as metal-base transistors, permeable-base transistors, and multi-layered-quantum-well and superlattice devices. MBE can be used to grow films of superior quality that no other method can match and therefore has attracted a lot of attention in research laboratories.

3.6 PROGRESS IN HIGH RESOLUTION LITHOGRAPHY

Advances in circuit design and new developments in processing technology were responsible for the major past growth of the IC industry. The current move to VLSI is based mainly on advances in lithography: achieving high resolution in the imaged resist, obtaining nearly exact mask-to-mask alignment, and providing faithful pattern transfer during the etching step. The availability of high quality masks with excellent dimensional control and clean processing (clean rooms, clean people, clean chemicals, clean water, etc.) has greatly reduced defect densities. These very complex and costly efforts have been necessary to allow the move to VLSI. It is interesting to notice how drastically the cost, the difficulty of wafer fabrication, and the size of the equipment must all continuously increase in order to continue to provide smaller new devices. It is also amazing how the performance of the systems that are made with these advanced technology ICs continues to increase, and yet the costs of these systems continues to decrease. This is a very good situation for the consumer.

In 1961, the feature size in the IC industry was approximately 25 μm. This was reduced to 5 μm in 1975. The feature sizes, as shown in Figure 3-4, used in high-volume production in 1984 were 3 μm, and smaller feature sizes are rapidly appearing. Current DRAMs, SRAMs, and 32-bit MPUs are using 1- to 1.5-μm processes, but the mainstream volume products are expected to stay at around 2 μm.

There is some confusion in the IC industry because some people refer to the *drawn* feature size as it exists on the mask and some refer to the *effective* feature size that includes the shortening effects that result from the lateral diffusions of the source and drain regions. We will use the *drawn* feature size.

The ultimate limit for practical channel lengths has been predicted

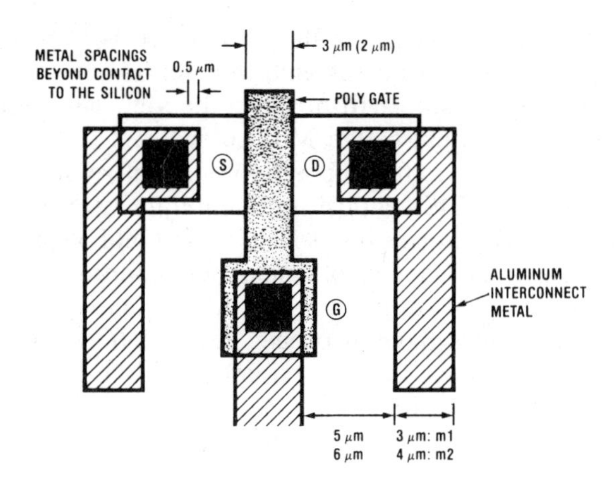

Fig. 3-4 The As-Drawn Feature Size in a MOS Layout

at approximately 0.1 μm, but other linewidths on wafers are anticipated to be smaller than 0.1 μm. For a relative size comparison, Table 3-1 lists the size, in micrometers, of a few selected items.

The extremely small sizes used in VLSI circuits greatly increases the requirements for cleanliness in all aspects of processing. Particles that were not "seen" in the older processes now create defects in the VLSI wafers. The capability to economically produce large die with complex circuitry will primarily depend on the ability to reduce the defect density in the silicon wafers.

Table 3-1. Using the Micro-meter (μm) Scale to Measure Other Items

Feature Size (μm)	Item
0.0005	Atom Spacing in Si Crystal
0.01	Polio Virus
0.2	Smallest Bacteria
1 (10,000 Å)	Yeast Cell
6	Red Blood Cell
100 (4 mils, 4×10^{-3} inches)	Diameter of Human Hair

Problems with the Mask Plate Material

The composition of the plates used for the masks has had to be changed. The earlier, popular, low-cost glass plates have a thermal coefficient of expansion that can cause a 2-μm runout on a 5-inch plate. Quartz plates have ultra-low thermal expansion and can provide less than 0.2-μm runout over a 6-inch plate for a temperature change of several degrees.

The flatness of the plates is also quite critical and is continually being improved. It is also interesting that the glass mask plates must be made thicker as they get larger to reduce image distortions due to *gravitational sag* that results from their horizontal position in the mask holders.

A Look at Wafer Imaging Techniques

The move to scaled NMOS memory production forced the IC industry to abandon the old standard emulsion masks. These were built by photoprinting on glass plates. Contact between the emulsion side of these plates and the wafer surface was used to transfer the image from the mask onto the wafer surface. This is known as contact printing.

The dimensional stability of the emulsion film was not adequate to pattern small features. The direct wafer contact also caused not only mask damage but also undesirably transferred photoresist from wafer-to-mask and mask-to-wafer. The soft, easily damaged emulsion film was replaced with a more dimensionally stable and tougher thin metal layer, such as nonreflective chrome or iron oxide, which was placed on the glass masking plates. It was then discovered that the previous soft emulsion had been cushioning stray particles and the smaller sized epi spikes (sharp vertical projections from the surface of an epi layer), so other masking techniques were needed.

The problems with contact printing were solved by new masking equipment. Proximity printing (near contact, but not projection), the more mature optical 1:1 projection printing, and optical wafer stepping (10:1 reduction or 1:1) have been introduced. The optical 1:1 projection aligner is an ingenious mechanism that makes use of first-surface mirrors to avoid the distortions and high costs of complex lens systems.

Diffraction and standing wave effects, resulting from the relatively long wavelength of the exposing radiation, was thought to limit optical lithography (photolithography) to about 2-μm feature sizes. This increased the interest in shorter wavelengths, such as ultraviolet (3000-Å or 0.3-μm wavelength: $1Å = 10^{-10}$m and $1\ \mu m = 10^{-6}$m) and deep ultraviolet (2500-Å wavelength). Researchers currently expect the optical lithography will be extended to provide 0.35- to 0.5-μm features in the future.

Problems exist when shorter wavelengths of radiation are attempted because the glass plates become strong absorbers at wavelengths less than 3500Å. With radiation of 3100Å and shorter, quartz masks must be used. Wavelengths shorter than deep ultraviolet create special problems because of heavy absorption by the quartz mask plates. This forces a major change in the radiation source because x-rays (10Å to 150Å wavelength) must then be used because this is the next range of wavelengths for which masks can be made. In addition to x-rays, electron beams and ion beams are being considered for IC wafer lithography.

To produce small features (2 μm and smaller) the problem of silicon wafer size distortion had to be accounted for. A silicon wafer, simply heated and then returned to room temperature, would present no wafer size distortion. But when an oxide layer is grown on this wafer surface, problems exist because the thermal expansion coefficient of this oxide layer is approximately one-fifth that of the underlying silicon. This creates an effect similar to a *bimetallic strip* and causes a convex wafer on the SiO_2 side to result. This size distortion is further complicated by the location and sizes of the windows that are etched in this oxide layer, because this complex oxide pattern directly affects the resulting wafer size distortion.

As a result of this dimensional instability, only relatively small areas must be individually aligned and exposed across the wafer surface. These areas may contain multi-die, although groupings of less than 10 die are usually used. This smaller area die patterning directly on the wafers (*Direct Stepped Wafers*, DSW) reduces the number of wafers that can be processed per hour, therefore increasing the fabrication costs.

Optical projection aligners, used for 5- to 3-μm feature sizes, must be replaced by optical wafer steppers (DSW) at 2-μm and smaller feature size, due both to wafer size distortion and the need for the more precise alignment capability that is achieved with the steppers. These steppers are expected to reach a limit at 0.35- to 0.5-μm feature size; therefore, to go below this requires e-beam, and eventually an x-ray or ion beam exposure system.

The major limitation is not only in the resolution of a given lithograhic system, but also in the mask-to-mask alignment accuracy that can be obtained. Linewidth patterns as small as 0.75 μm have been resolved and etched using 3100Å wavelength radiation. A rule of thumb is that the alignment precision should be 20% of the aluminum feature size. Wafer steppers which align through the reticle and through the lens can achieve die alignment accuracies near $\pm$ 0.1 μm consistently across the wafer, yet the processed silicon wafer tends to be stable to 0.1 μm only over very small areas, regions of the order of a few cm^2. Low temperature processes are used to help reduce this wafer size distortion problem.

Various dimensional tolerances are monitored in the masking and etching steps. Critical Dimensions, CDs, are indicated that the masks must produce with a close tolerance. All the dimensions associated with the masking of an IC are not of equal importance. For example, the channel length (the width of a defining poly layer) is one of the most critical dimensions. A set of design rules indicates the tolerances that exist for each dimension.

At small feature size, the problem of silicon defects creates heavy yield loss. Every processing step adds to the total number of defects. Noncontact mask aligners help insure that the masking operation does not significantly add to the bulk silicon defect density. For products that are in high volume production, such as memories, the masks are also individually repaired and returned to service if the wafer sort tester shows a consistent mask-related problem in a given die location.

E-Beam Exposure Systems

E-beam Exposure Systems (EBES) are presently used to manufacture IC masks. EBES can produce hard-surface masks, chrome or iron oxide, with better resolution and line width control, and lower defect density, than the older emulsion masks that were produced by optical mask-making systems.

EBES are now also used to scan and expose a special e-beam-sensitive resist layer that is placed on each wafer. The alignment precision of these systems is 0.1 μm. Beam reflections off of alignment marks are used. These marks are usually placed at the four corners of each die, within the scribe grid pattern of the wafer. The resolution of an EBES is 0.5 μm. Positioning of the wafer table can be measured interferometrically to 0.03 μm! These e-beam wafer-imaging systems do not require the use of masks and therefore can be used to expedite the first silicon for a new product, but are high in cost and low in throughput. EBES are expected to take over when the industry pushes feature sizes below what can be achieved with optical systems.

Advanced techniques considered for e-beam exposure systems propose a large area aperture and even variable aperture sizes to more rapidly *paint* large areas rather than to repetitively scan with the minimum spot size, 0.5 μm. Also considered are e-beam imaging systems where e-beam scanners are used to produce a high resolution, first layer mask in a multilevel masking technique. Backscattering of the electrons from the substrate has limited the single-layer exposed resist images to relatively low height-to-width ratios; ratios greater than 1:1 are difficult. This problem has been overcome in multi-layer resist techniques (which we discussed earlier in this chapter) where the e-beam is used to define a thin top layer.

X-Ray Lithography

X-ray lithography was introduced in 1972 by Spears and Smith of the Lincoln Labs of the Massachusettes Institute of Technology. Replication of masks down to 500 Å (0.05 μm) can be easily obtained. One additional benefit of x-ray lithography is that *dust becomes transparent* and is therefore no longer a problem. A direct wafer stepper mechanism must still be used because of wafer size distortion to precisely align the x-ray-transparent mask that is held in proximity to the wafer surface.

A source of *bright* or *high flux* x-rays in a reasonably compact size is a problem because the present limited flux of the x-ray sources increases the exposure time. Development work also continues to improve mask stability, resist sensitivity, registration, and to increase throughput.

Ion-Beam Lithography

Ion-beam lithography is an offshoot of the x-ray system and also makes use of exposure through a mask. This idea was pioneered by Rensch, and his coworkers, at Hughes. The ion beam can be easily collimated (as compared to x-rays where a distant source must be used) and is composed of 200 keV protons at a beam current of approximately 1 μA.

The resist is considerably more sensitive to these more massive protons as compared with the small mass of the electrons of an e-beam system; a factor of 1845:1! This rapidly exposes the resist layer and raises the throughput, but this higher energy bombardment also heats the mask and this is suspected of causing pattern runout problems.

Both x-rays and ion-beam systems eliminate the diffraction problem which limits the resolution of photolithography. Scattering of the protons in the transparent mask material establishes the ultimate resolution limit of ion-beam lithography.

The $10K contact printing lithography systems of the 1960s are evolving into the $3M e-beam or $10M synchrotron (x-ray) systems of the late 1990s.

Control of Contamination

Silicon wafer fabrication has been continually plagued by many types of contamination. In lithography, the presence of dust and other particulate matter reduces yields. It has been determined that, following a smoking break, clean room operators can give off as many as 2 million particles per hour from their bodies, in contrast to approximately only 10 thousand particles per hour from a nonsmoker.

To help reduce particulate contamination, clean rooms with air showers, tacky pads, and laminar air flow from high efficiency filters are used. Operators that work in these clean rooms must put on special clothing before entering the clean room to restrict the entry of both human-borne and human-sourced contaminants that are now held contained by the clean room garments. Routine inspections of the clean rooms are made at various locations using particle counters that report the density of particles in the air that exceed a prespecified size. For example, a "Class 10 clean room" (or a "Class 100 clean room") means that only 10 (or 100) particles greater than 0.5 μm exist per cubic foot of air per minute of flow through the optical particle counter. Special clean-up procedures are immediately initiated if the particle count becomes excessive.

As the dimensions of ICs continue to be reduced, this particulate problem will become even more severe. Particles of only 0.2 to 0.25 μm in diameter (and eventually < 0.1 μm) will become important and the large chip sizes of the future will require reductions in the permitted defect density. This represents a major problem because process cleanliness will have to be improved by very large factors (two to three orders of magnitude!) and particle monitoring equipment will have to be improved in sensitivity, accuracy, measurement consistency, and throughput.

An innovation that eliminates the dust problem with mask-plates provides a transparent protective cover for the mask. This cover is held sufficiently off the imaged surface such that any dust that may adhere to the outer surface of this protective cover will be defocused and therefore not be printed on the silicon wafers. This *pellicle technology* requires modifications to the stepper machines and adds to the cost of each mask.

To help determine the cause of contamination in wafer fabrication, many sophisticated detection and analysis techniques have been used. A very effective tool is the Scanning Electron Microscope (SEM). The SEM uses an electron beam to produce secondary electrons on impact with a target. These secondary electrons are then used to produce the detected image. This has aided in identifying photoresist residues, observing cracks and poor step coverage in thin films, determining the presence of contaminating particulates, and also indicating pits and scratches on surfaces.

As the geometries are reduced, the SEM has become necessary to determine the pattern size, material quality, and the coverage of thin films over the surface topology. Optical microscopes are limited to approximately 1000X magnification and 10,000 to 60,000X are commonly needed in VLSI circuit processing; therefore, a SEM is now required. This equipment can be fitted with an x-ray analysis unit (Electron Diffraction X-ray, EDX) that will also identify some of the elements present in a contaminate. The resolution of the SEM can be 50 to 100 Å. As we will see in the next chapter, the SEM is also being used to "see" voltages on IC chips in a circuit troubleshooting application.

Transmission Electron Microscopy (TEM) systems are used for surface mapping of very delicate structures. TEM is not a production tool. A replica of the surface is made with a surface conformable plastic. This plastic replica is then peeled off of the sample and analyzed with the TEM.

A relatively recent application for wafer investigation, although not a production tool, is Auger (pronounced "*Oh Jay*") analysis. Scanning Auger Microscopy (SAM), one of many types of surface science equipment, is a microbeam analytical technique that is used to characterize the elemental and chemical composition of surfaces and interfaces. All elements, except hydrogen and helium, that are present in detectable amounts in areas less than 0.2 μm in diameter in the outer few atomic layers of a material can be identified with SAM.

The specimen must be placed in ultrahigh vacuum, which increases the time needed for an analysis. Incident electrons from the scanning microbeam collide with inner shell electrons of the atoms of the specimen. This creates an electron vacancy within these atoms. When these atoms return to their normal state, x-ray photons or additional "Auger electrons" are emitted. The kinetic energies of these Auger electrons is measured to identify the constituent elements of the specimen.

SAM is also used to determine the relative concentrations of various elements within the bulk regions of a silicon specimen as a function of the depth from the surface. In this case, successive surface layers are removed (etched) before each analysis and both operations are sequentially performed on the specimen while it is in the same piece of equipment.

An almost unbelievable tool is the Scanning Tunneling Microscope (STM) that brought a 1986 Nobel Prize to IBM physicists Gerd Binnig and Heinrich Rohrer. This instrument uses an atomically sharp probe that is moved back and forth over a surface at a distance of a few atomic diameters and provides images of individual surface atoms and the bonds that hold them in place. This allows a new way to gain understanding of how silicon atoms are arranged.

3.7 SOLVING THE INTERCONNECT PROBLEM

The intrinsic logic-circuit delay-times will not limit the VLSI performance. The speed limit will come from the interconnect parasitic capacitance. For example, the effect of halving feature size and doubling die size has been shown to *increase the interconnect delay by 16-fold*! Extra interconnect layers are appearing to help solve this problem. A second level of interconnect wiring can reduce the stray capacitance to the substrate to only 30% of that of the first layer. This improves speed by 3:1. A third layer of interconnect wiring will have approximately only 20% of the capacitance of that of the first layer. The use of the third layer of interconnect rather than the first layer will improve the speed by a factor of 5:1.

Multi-Layers

Three and four interconnect layer processes use two conductive layers for the circuit interconnection and one or two conductive layers for power distribution. This improves performance and greatly simplifies the automation software that is now doing the IC chip layouts. Without multilayer interconnects, VLSI semi-custom chips would be unacceptably inefficient in terms of both silicon area and circuit performance.

In the past, sometimes as little as 30% of the chip area was used for logic components, most of the chip area was consumed by interconnect metal and cross-unders. Area-consuming cross-unders can usually be avoided by use of two layers of interconnect. In some cases this can reduce chip area by as much as 50%! Another 50% reduction in chip area can many times be achieved by use of a three layer interconnect process. The first layer interconnects the individual logic circuits, the second layer is used for all the intraconnect traces that hook up these logic gates and run from the top to the bottom of the layout, and the third layer takes care of all of the intraconnect traces that run from left to right on the layout.

In processes that use three or more layers of metal, special steps must be taken to eliminate bumpy oxide layers and to round the edges of the via holes to achieve good metal coverage between the layers. For example, one of the many techniques to reduce the surface irregularities in the oxide dielectric layers used between the metal layers is shown in Figure 3-5. Note that a significant number of extra processing steps must be added.

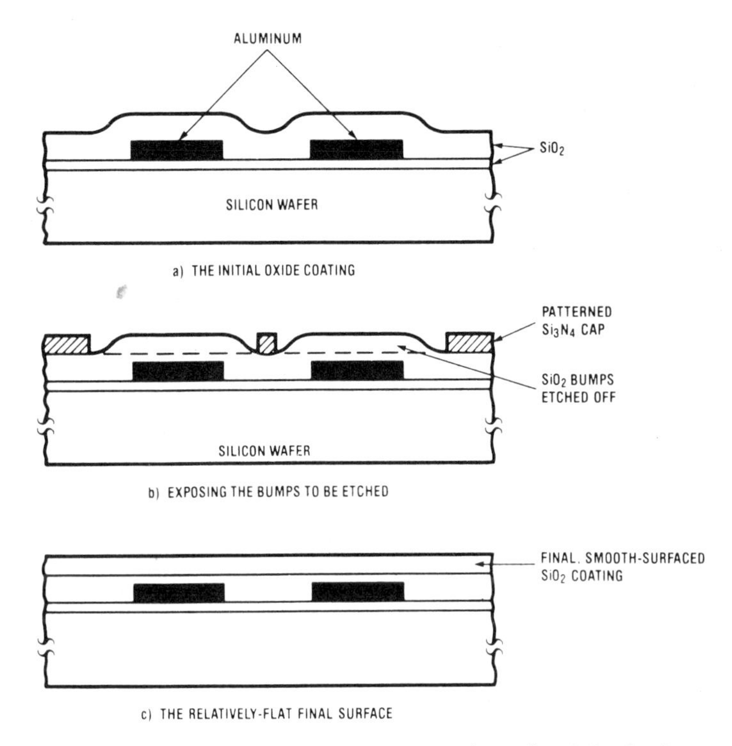

Fig. 3-5 Added Processing Steps to Provide a Flat SiO_2 Surface

Effects on Propagation Delay

To predict the *propagation delay time*, t_{pd}, that is associated with an interconnect line, we can use a lumped RC model and find

$$t_{pd} \cong 2.3RC$$

This makes use of a *short line* assumption and is valid for the circuitry of today.

As the spacing between the interconnect lines decreases below 1 μm, the parasitic capacitance between the *vertical sides* of the adjacent lines become significant in relation to the other interconnect capacitances. For separations less than 0.5 μm, this capacitance dominates the response time of the interconnect lines and imposes an upper limit on the frequency response of the advanced VLSI technologies.

Problems with Pure Metals

The use of pure metals, such as tungsten or aluminum, for added interconnect layers requires the elimination of all high temperature processing steps following metal deposition. Also, since neither metal can be effectively thermally oxidized, deposited dielectrics are required for the interlayer insulation. These factors tend to favor silicides for interconnects in memories (where the additional benefits of silicides can provide high dielectric constants for capacitors) and in microprocessors. Silicides will be considered in the next section. Multi-layer aluminum processes are generally used for gate arrays because the low resistance of aluminum reduces the series resistance of the longer intraconnect runs that typically exist in these layouts.

Although aluminum is a good conductor, it has several problems that must be accounted for when it is used in the manufacture of ICs. One of these problems exists when aluminum is in contact with silicon. The aluminum can migrate into the silicon lattice and will spike through, creating damaging electrical shorts when shallow junction depths are contacted. A solution to this problem has been to blend in 0.7 to 1.5% silicon in the aluminum film that is deposited. This satisfies the tendency for movement of aluminum atoms across the Al-Si boundary and therefore reduces this junction spiking problem.

Metal migration is the second problem with aluminum. This is the bulk movement of aluminum that occurs because of momentum exchange between conducting electrons and aluminum atoms. The *electronic wind* of conducting electrons will impact aluminum atoms that have been temporarily freed (because of thermal energy) from their positions in the aluminum structure. Atoms of aluminum will physically move in the direction of the electron flow. This relocation can cause voids to appear across an aluminum film which will interrupt current flow and cause circuit failure. The relocated aluminum atoms can simply be deposited in random piles, *hillocks*, or clumps of very strong single crystal "whiskers" of aluminum can grow up from the aluminum surface. These whiskers grow like a bamboo shoot. The new material that is added at the bottom pushes the "growing" rod up as if it were a living plant.

Elevated temperatures exponentially increase the magnitude of the metal migration problem. This failure mechanism also increases as the density of the dc current that is flowing in the aluminum film increases. For high reliability, the dc current density in an aluminum conductor should be limited to 2×10^5 A/cm^2. A blend of 4% copper is used in the deposited aluminum film to increase this current density. Also, traces of nickel (approximately 700 ppm) have been found to increase the mean-time-before-failure by a factor of 10. Electromigration continues to account for a large fraction of the field failures of ICs.

The problem of step coverage is aggravated with small feature sizes. The high current density that results over steps has forced the consideration of refractory metals, such as tungsten and molybdenum, to eliminate the electromigration problem and CVD processes to improve step coverage.

Polysilicon and Polycides

The polysilicon (poly) material that originally was used to provide the self-aligning silicon gate technology is often also used in multiple layers as interconnect lines. Some of the modern MOS IC processes use as many as three or four poly layers to ease the circuit interconnect problem.

Poly intraconnect lines that are hundreds of mils long can cause propagation delays in the μsec range because of the relatively large series resistance of the poly line and the stray capacitance of this line to the substrate. To reduce the resistance of these poly interconnect layers, metal silicides are deposited on top of a thin (0.2 μm) poly layer to form a *polycide* interconnect layer. This use of polycide can reduce the sheet resistance of the poly layer from approximately 15 Ω/□ to 2 Ω/□, but the even lower sheet resistance of an aluminum conductor, 15 mΩ/□, is needed for long interconnect runs.

It is unfortunate that the high value of resistance limits the usefulness of poly interconnect lines in the newer high speed circuits. Poly has many advantages: it is compatible with silicon processing, forms the self-aligning gate electrode in MOS transistors, provides stable threshold voltages for MOS devices, is relatively simple to deposit, is easily formed into narrow line widths, and poly easily forms an ohmic contact with both silicon and metal.

Another use for poly layers is being reported by researchers. This interesting work modifies the polycrystalline structure and converts it to monocrystalline silicon. Poly layers that were originally deposited onto oxide layers can be converted to single crystal. Therefore, transistors can be fabricated within these electrically isolated, stacked, vertical layers.

A possibility has been reported where the two complementary transistors that are used to form a CMOS logic inverter are made in a vertical stack with a single common gate at the midpoint, as shown in Figure 3-6. This idea of this *joint* usage of a single gate to drive both the N- and

P-channel devices led D. Kleitman to suggest the term JMOS (also JCMOS is used) to describe this structure. It is interesting that only one transistor type resides within the substrate, so the SCR problem also disappears. The relatively high threshold voltages of these transistors is one limit to widespread usage, but this is hoped to be a viable process in the future.

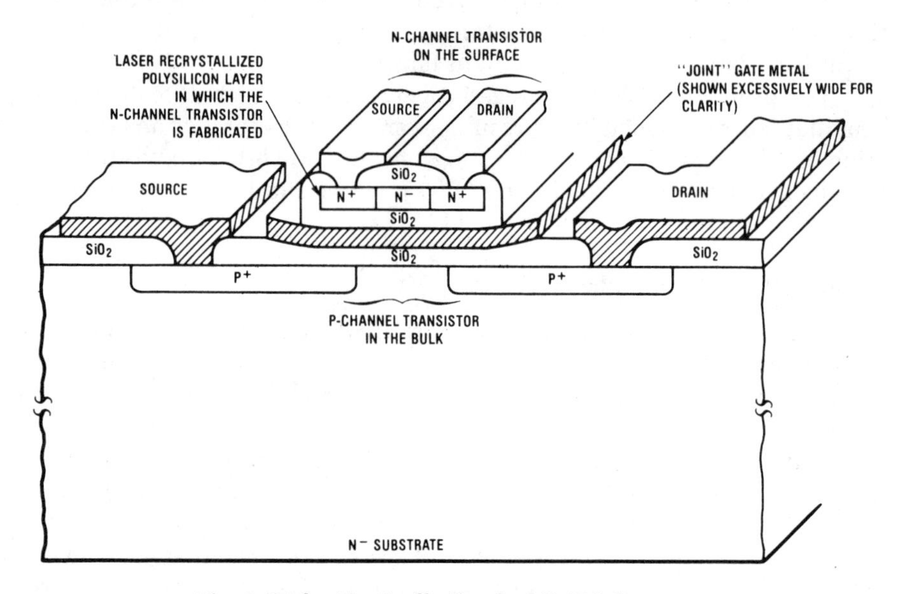

Fig. 3-6 The Vertically Stacked JMOS Structure

Laser beams have been used to convert a poly layer to single crystal silicon. Three basic ways are used to provide the desired crystal orientation reference for the newly formed film, as shown in Figure 3-7. Figure 3-7a uses the crystal structure of the bulk silicon, which is periodically exposed, to provide the crystallographic reference for the new film in

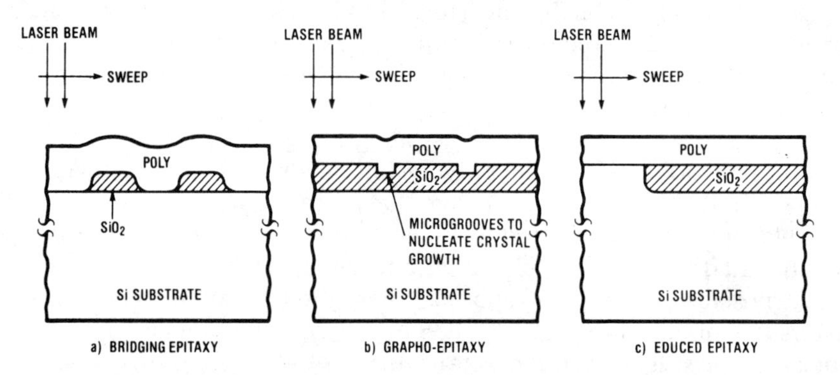

Fig. 3-7 Laser Conversion of Poly to Single Crystal Silicon

what is called *bridging epitaxy*. The *grapho-epitaxy* technique of Figure 3-7b makes use of *micro grooves* that are periodically etched into the surface of the oxide layer to nucleate single crystal formation of the silicon film. The *educed* (to bring forth or draw out) *epitaxy* of Figure 3-7c is similar to the bridging epitaxy of Figure 3-7a.

Connecting to the Outside World

A severe interconnection problem involves connections from the small sized IC chip to the "heavy cabling" of the outside world. The propagation delay increases approximately an order of magnitude for these off-chip connections as a result of the much larger stray capacitances which must be driven. This increases the importance of proper partitioning of the new VLSI systems that minimize chip-to-chip connections.

CHAPTER IV

The Present CMOS VLSI Revolution

People in the electronics industry are talking about Very Large Scale Integrated (VLSI) circuits, yet many may not be aware of all of the resulting effects of this electronic revolution. In this section, we will provide some insight into the many changes that may be brought about by the move to VLSI.

The complexity of ICs has approximately doubled every year since their introduction. Cost per circuit function has decreased several thousand-fold, and system performance and reliability have both improved dramatically.

As we moved toward more complex chips, problems resulted from excessive power dissipation and large magnitudes of dc current density in the interconnect metallization that existed with NMOS products. This forced the semiconductor industry to change to power-saving CMOS technologies. This renewed interest in CMOS has caused the older metal gate CMOS processes to rapidly evolve to the present advanced state of the art in silicon processing. Today, CMOS products have high-speed performance and are overtaking NMOS products. But bipolar processes are also rapidly advancing and will therefore continue to retain approximately a 2:1 performance advantage over MOS products. In addition, CMOS easily allows the mixed analog and digital circuitry that is often needed for single chip semicustom and custom ICs.

The increased complexity of advanced CMOS processes has caused confusion within the industry. Processes with multiple poly or multiple metal interconnect layers exist. Some processes are known as *standard-well* and some as *reverse-well*. All of these processing combinations are provided with feature sizes, the smallest resolved pattern on the die, that steadily decrease with time. High-volume production is expected to eventually end up at the submicron level.

The circuits that can now be placed on a single chip can utilize up to a few million transistors. Chips of the future are expected to hold up to 10,000,000 transistors as we become able to process 0.1-μm linewidths. These chips will be the new *Integrated Systems*, or ISs. Even today, many complete electronic systems can very likely be placed on only a few chips.

This has created a strong interest in optimizing system architectures because the circuit complexity that can be economically placed on a silicon chip, including any and all forms of memory, has exceeded the conceptual abilities of today's system designers. We have entered an era in which we can do *more* than we know what to do *with*.

Fortunately, the complexities of VLSI design and mask layout are handled by computers. Software packages greatly reduce both the labor and the time needed to complete a new VLSI chip design. The modern attitude of the system designers is that an IC fabrication line is simply an output device for their computer-aided design (CAD) system.

The amazing thing is that the processing of silicon—one of the most exacting industrial processes—continues to allow increasing complexities of electronic functions at decreasing costs. Even today, there is a large increase in the electronic sophistication that is being added to help run factories and businesses. Complex electronics has also appeared in games, household appliances, and automobiles. As VLSI circuits become part of the systems of the future, we can expect electronic sophistications that are substantially beyond what we have seen today. For example, the VLSI home computer of the future will provide the performance that is associated with a relatively large main-frame computer of today.

4.1 FROM COMPONENTS TO SYSTEMS

A relatively clear functional separation has existed between the systems houses and the semiconductor houses: one deals with systems and the other deals with components. This began to change with the appearance of LSI and microprocessors. Semiconductor houses are beginning to move into the systems business because the increasing complexity of the modern ICs allows larger chunks of a system to be placed on a single chip. To obtain the knowledge necessary to supply these more complex chips, the IC chip designers are now forced to take the viewpoint of a systems designer. In addition, system houses continue to acquire their own IC manufacturing capability.

4.2 SPECIFYING THE NEW CHIPS

Building complete systems on a chip, instead of the more fundamental circuits, greatly increases the design cost of each new IC product. The large commitment of resources that is required also reduces the total number of new programs that can be undertaken. A serious problem is the increased risk that is associated with these more costly efforts. It always becomes easier to miss the market as the complexities of a product increase, because the market may not agree with all of the design choices made. Thus it is also natural to want to add all of the possible features so that the product will not be lacking the ones that are later found to be important. This, therefore, makes it very important to properly specify these new chips, because the increase in complexity also increases the turnaround time, and costs, for major changes.

4.3 THE EFFECTS ON THE SYSTEM DESIGNERS

In the past, the move from transistors to ICs placed a large part of the circuit design effort within the semiconductor houses. Systems have been built by assembling IC packages on PC boards. Larger and larger blocks of circuitry have continued to be placed in a single IC package. Many times, the systems designers have complained that the semiconductor houses were not building the right ICs. The move to VLSI has changed all of this—the detailed design of the systems is once again returning to the system designers. With this comes a new degree of freedom; systems are no longer limited by the availability of particular ICs. The system designers can now get whatever they want or can imagine. It is up to them.

It may seem as if the system designers must rapidly become IC designers, which is an overwhelming thought to many. Fortunately, the complexity of VLSI design requires extensive computer assistance. These new IC designers have their choice: they can use standard *library* circuits or try their hand at the details of circuit design. In either event, the system designers will be free to try many new architectures.

The problem, now, is to determine ways of optimizing or intelligently choosing from the wide variety of architectural options that are possible. Theoretical people are busily at work to solve this problem. This increased flexibility is making the architectural design the most interesting and the most vital part of a new system. (We will consider new architectures in the last chapter.)

4.4 THE EFFECTS ON THE ELECTRONIC COMPANIES

The move to more expensive tooling always has both advantages and disadvantages to the electronics companies. If a company is large enough to afford the initial, relatively large investment, then its products can benefit from the production economies of electronic systems that contain very few semiconductor packages. Competition in electronic systems will therefore increasingly favor the large system houses as we move into the VLSI era.

4.5 THE EFFECTS ON THE SEMICONDUCTOR COMPANIES

Semiconductor companies are doing business in a new way. Working as a Silicon Foundry operation, the semiconductor company receives chip designs and/or actual masks from customers. The semiconductor company then fabricates wafers and packages the resulting circuits to the customer's specifications. One foundry may even look for wafer fab assistance from a different foundry: a *reverse* foundry operation.

It is true that the cost reductions in the new VLSI electronic systems will greatly increase the market demand for ICs; and, hopefully, at a greater rate than that at which the total number of packaged semiconductor products is declining. This conversion to a *smokestack industry* is the mark of the maturity of silicon processing. A new discovery or perhaps gallium-arsenide-based semiconductors will soon be needed to stimulate the IC industry. The present momentum of the semiconductor industry will last for many years. It is only the long term that is in question.

4.6 COMPUTER AIDS FOR VLSI DESIGN

Computers are necessary to *manage the complexity* that is introduced with VLSI chip design. To make the design task humanly manageable, many levels of abstraction must be used sequentially. The complex overall design task is broken up into a series of relatively easily-handled tasks. The desire is to have *extremely sophisticated computer programs* that are *extremely easy to use.*

This increase in complexity has caused fear among IC chip designers. They are terrified of being given back a large chip that is fabricated correctly, but doesn't work—and can't be diagnosed. Computers are used to eliminate human errors, the tedium, and the never-ending details that are associated with the design of the system, circuit, and chip layout.

The mask design (chip layout) effort has been a costly and a critical step in the design process. For this reason, semiconductor companies usually approach CAD from the *bottom up.* The new interest is to develop computer programs that will handle system design from the *top down.* These will provide increased aid for the system designers. For example, computers will handle the communication requirements of the design team and also high-level system simulations to simplify the choice of an optimum architecture for a new system.

Design tasks have been separated into *Functional Design*—the synthesis, verification, simulation, and testing that is done at the architectural, system, logic, circuit, device, and process levels—and *Physical Design*—the partitioning, layout, and topological analysis that takes place at all levels. The issues of functionality, testability, and the physical design factors must all be considered in parallel throughout the design process.

Hierarchical procedures are used that allow many details to be included in each step of the sequential design process. The idea is to pick up the details as you go through the design procedure, but to only do each operation once. For example, if a circuit cell is obtained from a library source (which takes the place of the SSI and MSI logic databooks), the detailed circuit analysis and the layout design rule checks should be part of the library data base and these details will not have to be recomputed. Also, if layout design rules are checked for each transistor as it is being used, then gates can be designed with these transistors without rechecking the layout design rules for transistors. Multi-gate blocks can also then be developed from these primitive prechecked gates and only the local gate interconnections need checking. In all cases, once a chunk of circuitry has completed layout and checking, a *protective field* is placed around this area of the chip in the computer's data base and all of the known-to-be-correct enclosed features, or feature groups, will not be disturbed or rechecked.

Finally, these multi-gate blocks can be interconnected and only the global block interconnections need checking. This hierarchical design rule check can speed up the checking process by a factor of two orders of magnitude or more when compared to the complex problem of checking all of the details that exist on each of the completed masking layers.

A further goal of the computer aids is to make use of a common data base so that the design can proceed smoothly with only a single entry of the desired logic or functional specification of the final chip. Unfortunately, this has taken programmers many years to accomplish because the early individual CAD programs were not standardized and the enormity of the individual programs that are used complicates this data base merging. For example, many of these programs involve hundreds of thousands of lines of code in a high level language.

Large memory resources are required and many hours of CPU time are involved in the solution of the problems associated with a detailed chip layout. The tasks of *automatic placement* of the cells and then the automatic hookup, *routing*, of these cells are enormous computer problems.

The creative contributions of the human mask designer are also extremely difficult for computers to emulate. The obvious way to solve a routing congestion problem may be very easy for an experienced mask designer to *see*, but computers are not yet this intelligent, so they must plod along through endless code and sometimes they, or more precisely, their programs, never can sort it out. For this reason, many of the computer aids have relied on human intervention to help solve some of the more difficult problems.

Cell and Functional-Block Libraries

Library cells and functional-blocks (such as SSI, MSI, ROM, RAM, PLA, and even a selection of useful linear and data acquisition circuits) are called out by computer programs to satisfy the given logic or system diagram that is to be designed as a VLSI chip. These cells maintain some degree of layout flexibility such that more compact chips will result. As simple as this sounds, it becomes difficult, for example, for a router program to accommodate cells with only two different heights. Further, the automatic router program will place constraints on the allowed layouts of these library cells. Finally, to initially put this library in place represents a considerable expenditure of resources. Much thought has to be given to determine what is really needed in the library. Many existing IC circuits may not be good library candidates because they haven't made use of structured design or regular layout approaches. (Regular layouts make extensive use of cells, such as programmed logic arrays, rather than random logic.)

Mask Design

The automatic mask design for a general VLSI chip is still a big problem. After the selection of an optimum collection of library circuits to realize the given logic or system function, these circuits have to be placed and routed. The optimum placement of these blocks is one of the hardest problems to solve and the global (overall) routing programs consume extremely large amounts of CPU time. In addition, special *compactor* programs are introduced that reduce or reconfigure the sizes of these building blocks to insure a more optimum use of the surface area of the silicon chip.

Hierarchical design rules, those aspects of the physical layout dictated by the requirements of the devices and the processing steps used in wafer fabrication, must also be checked by programs to insure conformance to all of the sizing and spacing specifications for each mask layer.

Place and route programs are necessary and may require intervention by a mask designer. In addition, system level simulation and modeling (including the characterization of the process at the transistor level) is needed along with system level testing and debugging tools.

Finally, the last step—handling the PC board layout for the system—is usually a part of the overall VLSI design package.

4.7 PROBLEMS IN CHECKING THE MASK DESIGN

To appreciate the problems in checking or doing a mask design, we will use an example to show the steadily increasing detail that can be placed on a single IC chip.

It is difficult for most people to comprehend the very small dimensions that are used to specify the minimum feature sizes on a chip. If we bring this into a more familiar set of measurements, the nature of the problem of checking or designing the masks can be more easily grasped. This dimensional transformation has been made by always using the more human-relatable measurement of ¼ inch to represent the minimum feature size.

In the late 1960s, the feature size was 15 to 20 μm so we would have to magnify these chips (15 μm features) by 423 to obtain our 1/4 inch *real world* feature size. The size of these magnified chips would be approximately 7 feet square (for a 200-mil by 200-mil die), and the details on these chips (with lines no closer than ¼ inch) were designed and checked manually.

The factor of two reduction in feature size that was possible by the early 1970s would increase our magnification to 800 and these enlarged chips would now occupy an area the size of the average living room floor.

Reductions in feature size by the late 1970s would require increasing our magnification to approximately 3200 and this chip complexity, still using ¼ inch for the minimum *real world* feature size, would cover the area of two tennis courts—still manually drawn and manually checked!

Continuing this to a 1.25 μm feature size would require a magnification of 5000 and these *real world normalized* masking layers would now cover two basketball courts. Think of how much drawing complexity can exist in this large area when the lines are are only ¼ inch apart! Here is where computer assistance comes to the rescue.

If we go to the submicron feature size of 0.5 μm, our required magnification becomes 12,500 and these masks would now cover two football fields!

Even with the relatively simple layouts of the past, the tedious job of checking mask designs has caused more than a few electronic design engineers, the author included, to wonder why they ever got involved with ICs. This also makes it more understandable why the manual design of LSI chips used to take a crew of mask designers many months to complete.

4.8 THE INCREASINGLY COMPLEX TESTING PROBLEM

The move to large-diameter wafers and complex circuits with long test programs has increased the total time that is needed to electrically test a finished wafer. Total test times per wafer can consume an entire eight-hour shift. In general, more circuits are also produced to keep up with the steadily increasing demand for ICs. All of these factors increase the number of wafer testers that are needed and the corresponding floor space that must be allocated to wafer testing.

The difficulties in testing digital circuits depend on the total number of states that must be tested, and the accessibility to the internal circuitry that is provided by the package pins. A figure of testing difficulty that has been used is *the package pin-to-gate ratio*. Products with 10 to 20 thousand gates in a 200-pin package simply do not have good *external access* to the *internal circuitry*. This makes testing extremely difficult and also very time consuming.

Further complications can be created by the architecture used. For example, a finite state machine, where state transitions depend upon the content of internal registers and internal flag states, can be very difficult to test. In many cases the architecture must be changed to increase the testability, or additional test pins must be added to increase the access to the internal circuitry.

A better solution to the testing problem is to incorporate built-in testing, self-testing circuitry. Only gross functional testing is expected to be done with the external tester and the detailed testing will be done on the chip. Some new chip designs are also making use of built-in redundancy and the capability for automatic reconfiguration to recover from internal fault conditions; for example, by using an on-chip E^2PROM, a memory that can be both electrically programmed and electrically erased.

As we moved into the VLSI era, the diagnostic testing of improperly functioning chips also has become a harder problem. In the past, internal access to troubleshoot the circuitry on an IC chip has been made with multiple metal probe-needles that were simply stuck into the aluminum metalization. The problems of excessive capacitive loading on the IC circuitry and the requirement for a larger number of probes has caused some researchers to investigate the use of a scanning electron beam, the Scanning Electron Microscope (SEM). (We discussed this in the last chapter.)

It has been know that static voltages on the surface of ICs can affect the secondary electrons that are emitted in a SEM. As may be expected, a positive voltage (logical "1") collects these secondary electrons and therefore produces a dark region on the display. A logical "0" (0 Volts) does not bother the secondary electron emission and therefore a light region results on the display. This *static voltage contrast* has also been extended to view dynamic voltages by pulsing very short (a few μs) bursts of the primary electron beam. This strobing is synchronized to the system clock, the video display, and also to the physical scanning of the IC surface. If the timing, or phasing, of the primary electron beam bursts is cyclically varied, "sampled" node voltages can be observed. Clock signals can, for example, be made to "appear to move," (as a sequence of dark and light intervals) from a clock generator out to the logic circuitry. Any open lines or other disturbances along this clock line can therefore be "seen" to interrupt this slowly propagating clock signal. This is similar to the stroboscopic lights that can stop, reduce the speed, or appear to reverse direction in mechanical systems. People who have seen actual demonstrations of this strobed SEM display are usually very impressed. It's almost a cartoon animation.

By using these SEM techniques, data from a large number of nodes can be easily obtained and the circuit disturbances, that the act of measuring causes, are reduced. These are diagnostic, not end-of-line testers. Similar to the metal probe, this only operates on exposed conductors—oxide layers must be removed. Multi-layered ICs need to bring up test points or *observation points* to the top interconnection layer.The eventual small geometries also present focusing, field of view, and alignment problems for this technique.

A further restraint of these SEM diagnostic systems is that all of the external electronic components required to establish a repetitive *program loop* with the IC circuit to facilitate the testing must be placed not only close to the IC chip, but also this complete breadboard must be located inside the vacuum chamber of the SEM. For this reason, the IC die is mounted in a ceramic package and is operated with the lid removed so the SEM can "see" the die surface.

A relatively recent SEM technique (suggested by H. P. Feverbaum) combines the stroboscopic voltage conrast with the sampling mode. The pulsed primary electron beam is synchronized with a repeating group of IC system clock pulses that establish a repeating program (or logic sequence) loop. The phase shift, ϕ, that is used between this program loop timing reference and the triggering of each narrow electron beam burst is continuously altered, but at a slow enough rate that a sufficient number of pulses are returned to provide a good quality (noise free) display. While this continuous phase shift cycle is taking place at a relatively high rate,

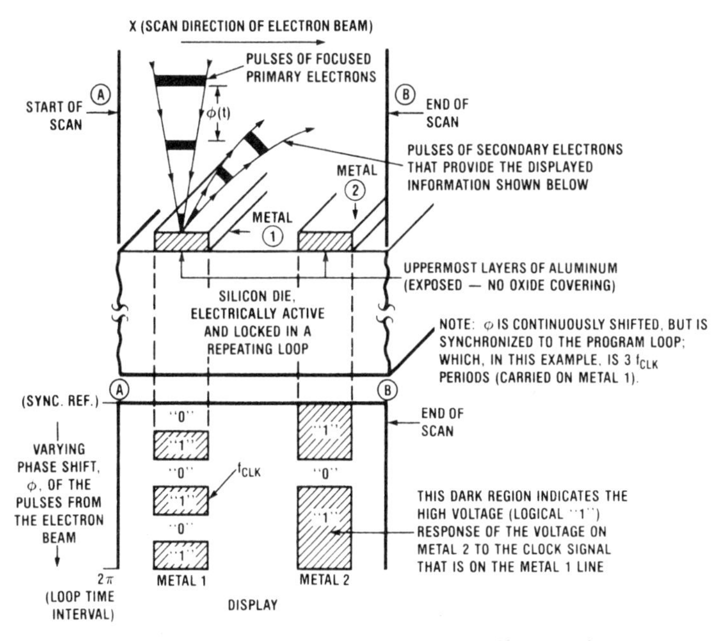

Fig. 4-1 Logic State Mapping to Diagnose Chip Performance

the electron beam is slowly swept across a repeatedly scanned surface of the IC chip. The voltages that exist at various times within the program loop (which now converts to particular values of ϕ) on the IC conductors are displayed. As in static voltage contrast, a logical "1" is dark and a logical "0" is light on the resulting display.

A diagram of this *logic state mapping* is shown in Figure 4-1. Unfortunately, in this illustration, the string of the light and dark spacings that correspond to the digital voltages on the two metal traces shown, have been placed so that time (ϕ) on the display runs vertically. This has been done so that the die surface scanning could be represented. The actual display would, of course, be rotated as shown in Figure 4-2, where the horizontal axis is the timing of the program loop. This logic-state mapping has also been modified to provide a *timing diagram* by sequentially aiming the electron beam at a given sequence of conductors. The results are now presented on a conventional logic analyzer to provide a familiar display format.

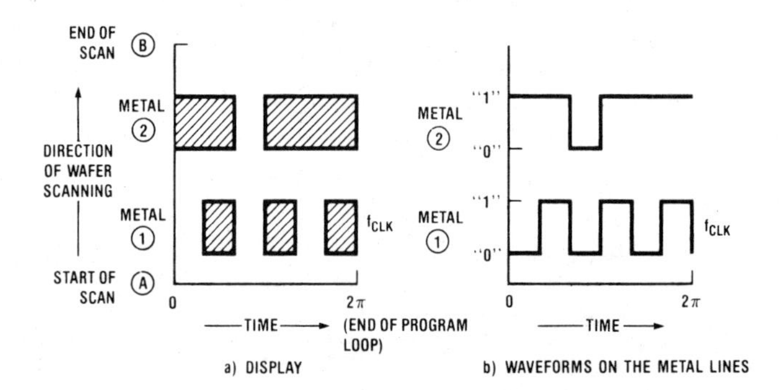

Fig. 4-2 Interpreting the Logic State Map Display

Speed limits of an IC can be localized if the proper program loop is set up. Asynchronous systems cannot be handled by these electron scanning diagnostic techniques, but it is expected that analog voltages can eventually be measured to an accuracy of ±10 mV!

CHAPTER V

Standard Circuits in the New CMOS Era

The requirement for standard circuits will continue for a long time. As we move into the CMOS VLSI era, the main changes that are occurring in digital circuits are: (1) the greatly increased functionality, (2) the increased package pin count, and (3) the reduction in the power consumed.

5.1 CMOS LINEAR CIRCUITS

The move to CMOS is having a major effect on the way linear circuits are designed. CMOS has the advantage over NMOS of providing complementary transistors which is very desirable for the linear circuit designers. The new thing is that the analog switch is available with CMOS. This is forcing linear circuit designers to look for ways to make optimum use of this new design element and is causing a *move to sampled-data circuits.*

There are basically three ways to build an electronic system: fully digital; fully analog where the input voltages are continuously applied and operational amplifiers are used—the analog computer approach; and sampled data, where "samples" are periodically taken of the input voltages. Sampled-data systems lie in between the completely digital system and the completely linear system.

As we have previously noted, the complexity of the CMOS process aids a linear circuit design by providing many additional devices. The availability of a zener diode and the parasitic vertical and lateral bipolar transistors are very useful components to add to the linear designer's bag of tricks.

A bipolar process has been combined with a CMOS process to end up with the best of both worlds. (It is interesting that a different type of a bi-CMOS process is also becoming very popular for digital circuitry.) The added costs for such a combined bipolar-CMOS process still favor a circuit or system solution that makes use of the processes available from the high-volume standard fabrication lines, but some interesting circuits, both linear and digital, can be built with this particular process combination.

Progress in CMOS Op Amps

The complementary transistors available with a CMOS process have allowed linear designers to essentially use existing bipolar circuit designs by replacing an NPN with an N-channel and a PNP with a P-channel MOS device. These early CMOS op amps were used on large custom IC chips and could outperform the NMOS op amps, but the all-around performance specs of the highly evolved bipolar op amps were hard to compete with.

There have been two parallel efforts in the design of CMOS op amps: continuous linear circuits and sampled data circuits. Most op amp users prefer the continuous or traditional op amp circuits because there are no noisy clock signals or sampling glitches to contend with. If these new noise sources can be allowed, the sampled-data circuits offer the benefits of essentially a zero offset voltage and also zero temperature drift in this offset voltage.

It was initially thought that the offset voltage of a continuous CMOS op amp would always be large. This offset voltage problem was helped in the layout of bipolar op amp chips when the linear designers made use of a cross-coupled quad of transistors to form the input differential pair (two transistors on each side). This has now been extended to a "checkerboard-coupled" hexadecimal of transistors (16 individual transistors—8 on each side) to reduce the offset voltage of a CMOS op amp. Ways are also being found to increase the open loop voltage gain that is obtained with these op amps. We must remember that the early bipolar op amps (which were built with the T^2L bipolar process) did not provide very impressive specs, and it took many years of evolution, and many new parasitic devices, to provide the high-performance bipolar op amps that we take for granted today. There is much hope for future improvements in the performance of CMOS op amps.

CMOS linear design is similar to CMOS digital design in that, once a basic circuit topology has been decided on, the major part of the design effort involves selecting the W/L values for all of the transistors. In linear bipolar design, standard, minimum geometry transistors were most often used and the detailed design procedure was to determine the values of the resistors that were used in the design.

An example of a second-generation CMOS op amp design is shown in Figure 5-1. A zener diode is used in the biasing circuitry. Notice that two parasitic NPN bipolar emitter followers are incorporated within the op amp design. The first, Q6, is used to provide a low impedance dc biasing voltage for the source of the N-channel transistor, Q5. This emitter follower, Q6, is used to improve the circuit symmetry which will reduce offset voltage because the drain-source biasing voltages for Q3 and Q4 are matched.

The second emitter follower, Q10, is used on the upper side of the

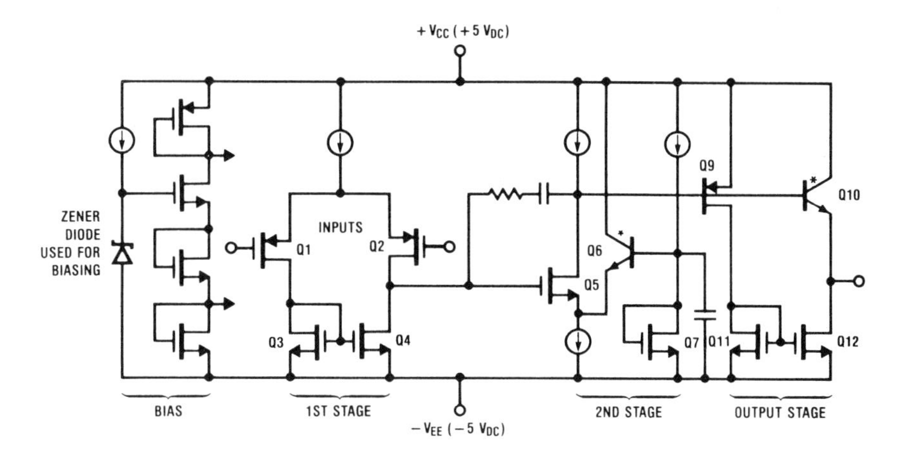

Fig. 5-1 A Second Generation CMOS Op Amp

output stage to allow sourcing current to an external load. The higher transconductance of a bipolar transistor, as compared with an MOS transistor, allows this transistor to be a relatively small geometry device. The performance specs of this op amp are:

PARAMETER	VALUE	UNITS
A_V (open loop voltage gain)	80	dB
f_u (unity gain frequency)	2.5	MHz
V_{OS} (offset voltage)	±7	mV
SR (slew rate)	3.5	V/μs
I_D (current drain)	1	mA

Newer circuit designs for CMOS op amps are using current-biasing tricks rather then voltage biasing and more than two stages of gain to improve the performance specs. One of the design considerations for these op amps is that they should make use of a standard digital CMOS process and not require any process changes just to support the op amps. Thus they would be available to be used as standard cells to provide linear functions on LSI chips.

The limited transconductance of the MOS transistors has forced the use of a three-stage op amp rather than the conventional two stages that have become common for bipolar IC op amps. An example of a "third-generation CMOS op amp" is the three-stage op amp shown in the block diagram of Figure 5-2. In addition to the frequency compensation capacitiance, C_c, two additional feedforward capacitors are used. The output

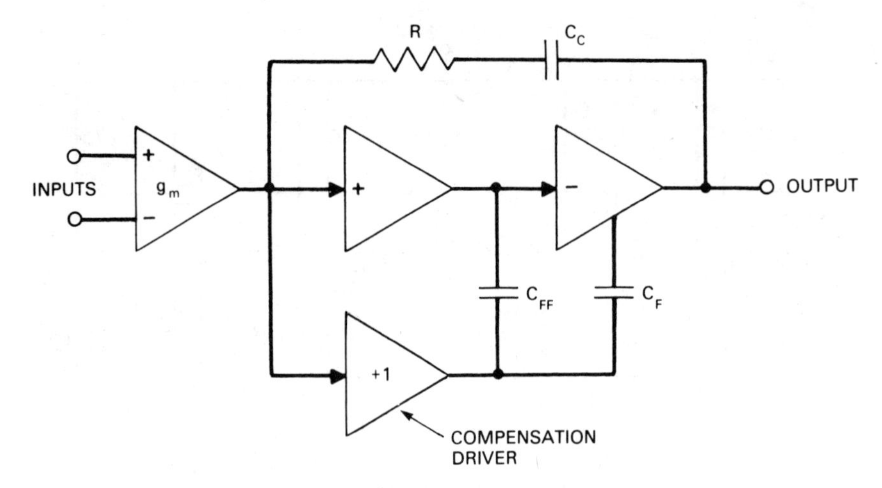

Figure 5-2 The Block Diagram of a High-Performance CMOS Op Amp

of this op amp is the same point as the output of the integrator; a separate unity-gain buffer stage is not used.

The previously mentioned "checkerboard-connected hexidecimal" of 16 total MOS transistors was used to solve the usual problem of high offset voltage in an MOS op amp. Native (unimplanted) P-channel transistors were used for this input stage to provide the lowest noise performance.

Both of the inputs to this CMOS op amp have on-chip protection circuitry that is similar to what is used on CMOS digital circuits: an input poly-silicon protection resistor and a diode from the internal side of this resistor to each supply. It is the reverse-biased leakage currents of these input protection diodes that cause this op amp to have a dc input current. In practice, the upper diode (to $+V_{CC}$) provides more leakage current than the lower diode (tied to $-V_{EE}$) and therefore the input current of this op amp comes out of the package (the direction you would expect for a PNP differential input stage). This current is typically only 50 fA at room temperature (less in a plastic package), varies with the input voltage at the rate of about 10 fA/V (an equivalent input resistance of 100 T Ω), and increases to only 5 pA at +125°C.

To assist the design of this CMOS op amp, use has been made of a parasitic lateral NPN transistor to provide controlled-magnitude biasing current references by using the circuit shown in Figure 5-3. The vertical collector of this NPN still exists and collects approximately two-thirds of the total emitter current.

This bias circuit makes use of a 4:1 emitter-area scaling between the differentially connected NPN transistors Q7 and Q6. This provides a difference in the V_{BE} voltages of these parasitic lateral NPN transistors (a Δ

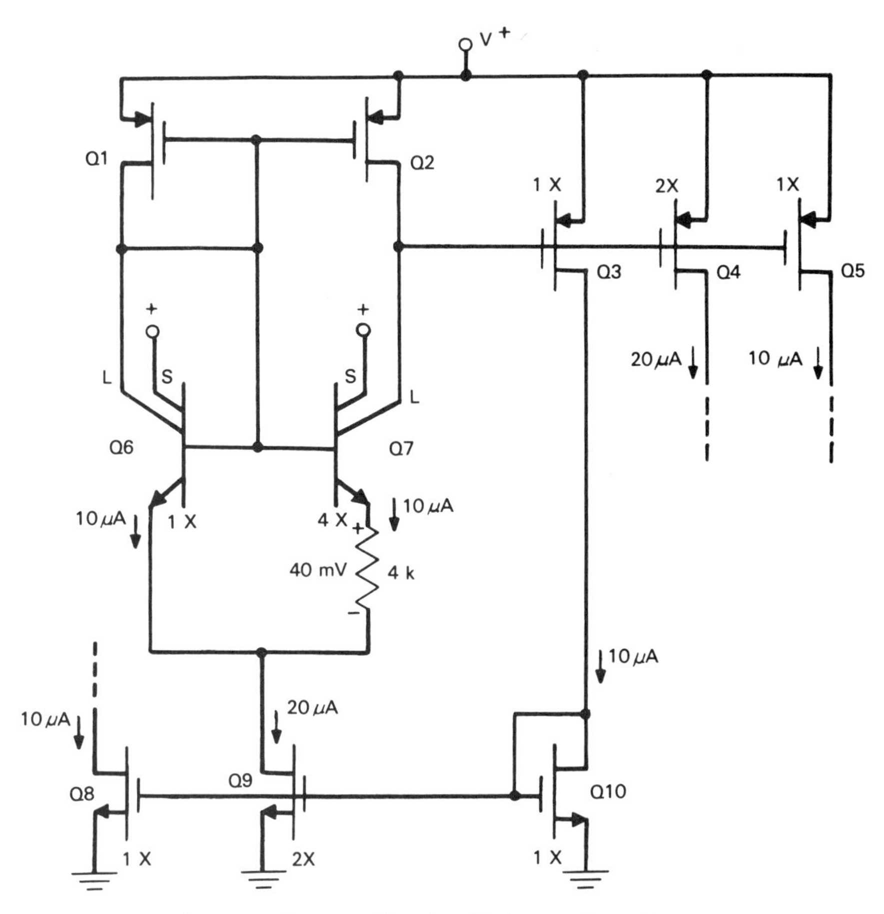

Figure 5-3 The Δ V_{BE} Biasing Circuit which uses Lateral NPN Transistors

V_{BE}) of 40 mV. In operation, with the NPN differential amplifier at balance, the 4-kΩ resistor in the emitter of Q7 has this 40 mV developed across it and therefore provides a 10-μA reference current flow.

As mentioned previously, in designing with MOS transistors the equivalent of emitter-area scaling is provided instead by scaling the W/Ls (the channel width to channel length ratios). Knowing that the total tail current of the NPN differential amplifier is 20 μA allows W/L scaling to be used between Q9 and Q8 to provide, for example, the proper gate-source voltage biasing such that Q8 will carry 10 μA of drain current. Similarily, Q10 is scaled to carry 10 μA, and the drain current of this transistor is used as a reference for transistors Q3, Q4, and Q5. In this figure, Q4 will supply 20 μA of drain current and Q5 will supply 10 μA. It is interesting that a familiar bipolar circuit (using a parasitic lateral NPN bipolar transistor) has allowed the design of a relatively predictable biasing circuit (this is not easily done with only CMOS devices) that will provide biasing currents for the rest of the CMOS op amp circuitry (via transistors such as Q4, Q5, and Q8—and any additional transistors and scalings that may be needed).

In this CMOS op amp, the complementary output MOS transistors are connected as common-source amplifiers. The output of the op amp is the common connection of the drains of these output transistors. This connection (collectors out) is rarely used in bipolar op amps but does provide the advantage of a large output voltage swing.

Another interesting aspect of this design is the way that the output short-circuit current is limited. In bipolar op amps, the output current is usually passed through a "sense resistor" just prior to coming out of the output lead. The voltage that is dropped across this sense resistor will eventually be of a large-enough magnitude (when the maximum amount of output current is reached) that special current-limiting circuitry, which is normally dormant, will come to life. This current-limiting circuitry then removes or steals base current from the output transistors and thereby provides an output current limiting function. The unusual thing with this CMOS op amp is that output current limiting is provided by simply limiting the magnitude of the gate-source drive voltage that is applied to the output FETs.

The overall performance on a 5V supply that has been achieved with these CMOS op amps is equal to or better than that of most commercially available bipolar, bi-FET, or CMOS quad amplifiers. Some of the key specs are:

PARAMETER	VALUE	UNITS
I_{IN}	50	fA
A_V	>100	dB
f_u	1.5	MHz
V_{OS}	± 1	mV
SR	1.7	V/μs
I_D	350	μA/OA
I_{OUT}	± 22	mA
V_{CC}	4.75 to 15.5	V_{DC}

This performance has been achieved by a combination of both circuit design and layout techniques, and therefore these op amps are suitable for incorporation on system chips that are built with the latest digital CMOS processing technology. (For more information on this op amp see: "A Quad CMOS Single-Supply Op Amp with Rail-to-Rail Output Swing," by Dennis Monticelli in the December, 1986, issue of the *IEEE Journal of Solid-State Circuits*.)

These amplifiers will provide competition for both the existing bipolar op amps and also the sampled-data CMOS op amps. Sampled-data CMOS op amps haven't yet captured a significant share of the op amp marketplace. As we will see in one of the following sections, the CMOS sampled-data voltage comparator, however, has become very useful for a line of CMOS analog-to-digital (A/D) converters.

Switched-Capacitor Filters

Switched-capacitor filters are an excellent example of the benefits of a CMOS sampled-data circuit approach and provide many significant performance advantages for these filters. These switched-capacitor filters are replacing the older RC active filters that were realized with resistors, capacitors, and op amps. Both of these circuit approaches to build filters replace the large and heavy inductors that were required for a frequency selective circuit, a filter circuit, with the smaller sized and less costly RC passive components and the low cost IC op amps.

To better understand the basic idea of these filters, we will first consider the more familiar multiple-input summing amplifier shown in Figure 5-4. Notice that we have labeled the (−) input, the *inverting input*, of the op amp as a *virtual ground*. This input node will always be at essentially ground potential because the very high gain of the op amp will allow any desired output voltage to exist with only an extremely small voltage at this inverting input. For example, if the op amp had an open loop voltage gain of 100,000 and an output voltage of 10 V_{DC} was required, the voltage at the inverting input needed to supply this output signal can be found as:

$$V_{IN}(-) = \frac{-10V}{100{,}000} = -100\ \mu V$$

and this 100 μV signal is small enough that it can usually be neglected. This keeps the inverting input of the op amp at *essentially ground* level and therefore has allowed this node to be called a virtual ground.

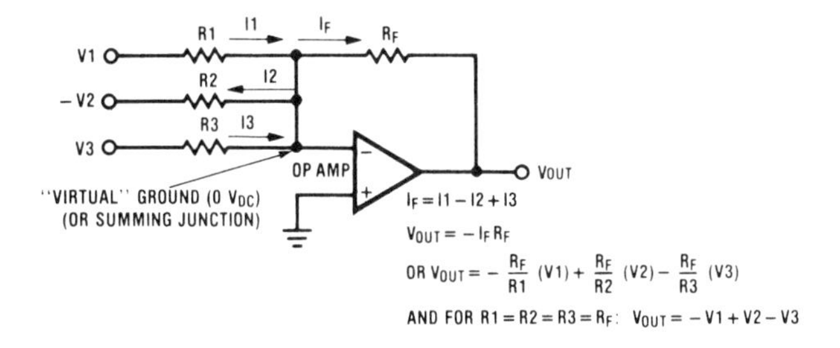

Fig. 5-4 The Multi-Input Summing Amplifier

With this inverting input node at essentially ground, notice that there will be practically no interaction between the multiple input signals. This is one of the advantages of this type of multi-input summing circuit.

This circuit is simply obeying Kirchoff's current law that states that the algebraic sum of all of the currents that enter a node must be zero. This means that the op amp must automatically adjust its output voltage to whatever is needed so the surplus of the input currents is always absorbed by the output of the op amp via the feedback resistor, R_F. Once this had been done, the resulting output voltage, V_{OUT}, will then equal the desired algebraic sum of the multiple input voltages.

The important thing to notice about this circuit is that *continuous input currents are compared*. When we change to a sampled-data circuit approach, *input charges will then be similarly compared*. Let's now see how this conversion to charge takes place in a switched-capacitor filter.

Many of the switched-capacitor filters have replicated (but in a sampled-data way) the very successful three-op amp *biquad* or *state variable filter* that uses the basic circuit that is shown in Figure 5-5.

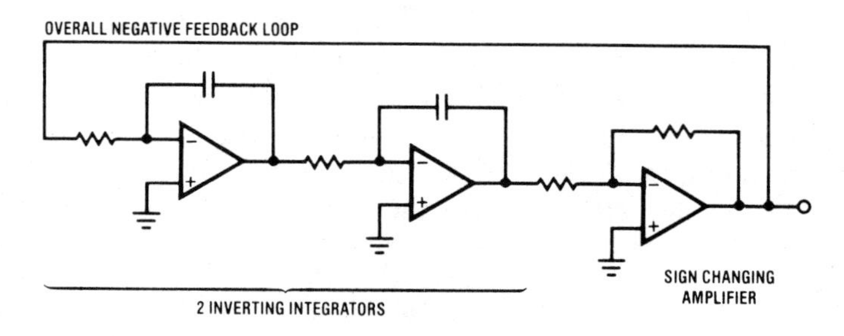

Fig. 5-5 The Successful RC Active Filter

The input and outputs of this circuit have been omitted to emphasize that the basic loop of this filter is composed of two integrators and a sign changing amplifier. Each of the integrators provides a polarity reversal. To allow tying these two integrators into an overall feedback loop, the sign changing amplifier is needed so this overall loop will provide negative feedback.

From the standpoint of RC active filter theory, each integrator provides a pole on the negative real axis of the complex frequency plane, the s-plane. The two negative real poles provided by the two integrators are then made to move off the negative real axis, because of the action of the overall negative feedback loop, and to assume the complex conjugate pole locations that are normally provided by an RLC network. This is the way an RC active filter simulates the performance of a passive LC filter network. (For those readers who find this explanation too sketchy, please refer to the author's previous book, *Intuitive Operational Amplifiers*, where these concepts are more completely developed.)

With a switched-capacitor filter, we must find a way to provide an integrator function, but in a sampled-data equivalent circuit. The details of this conversion to sampled data are indicated in Figure 5-6, where we see that the bipolar op amp is simply replaced by a CMOS op amp and the integration capacitor, C, is still used. Both the bipolar and CMOS are continuous op amps. The major change is to replace the input resistor, R, with a small-valued input capacitor, C_i, and to use a clocking signal, f_{clk}, to repeatedly switch this input capacitor from the input voltage, V_{IN}, to the summing junction of the op amp.

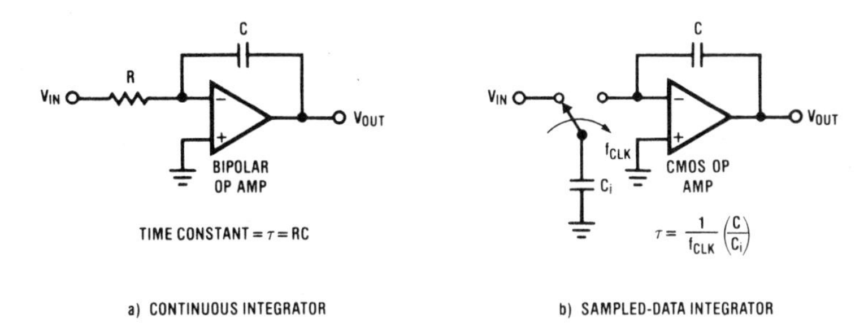

Fig. 5-6 Comparing Integrator Circuits

We replace the continuous flow of input current through R by a continuous stream of *clocked charge samples* that are acquired by the input capacitor, C_i.

Current is the flow of charge and is measured as coulombs per second. Therefore, if we *increase the clocking frequency* we will cause more current to flow and thereby simulate a *smaller value* of R. For this reason, the time constant of the switched-capacitor integrator is a function of the clocking, or sampling, frequency. This provides an easy way to *dynamically tune* a switched-capacitor filter under *computer control*.

We can now appreciate a few further benefits of the switched-capacitor filter. The time constant of the conventional integrator is the product of the R and C component values. It is the initial tolerances of these passive components that has required individual *tuning* of each integrator used in an RC active filter. Further, it is hard to keep this time constant independent of temperature changes because the temperature drift of both the R and the C are involved and it is relatively hard to match the temperature drift of an R with the temperature drift of a C.

Notice that the time constant of the switched-capacitor filter only depends on the *ratio* of two on-chip capacitors. This ratio can be supplied with both a low initial tolerance and a low temperature-drift. This time constant also depends upon the initial accuracy and the temperature stability of the clocking frequency. Crystal-controlled clocks can solve both of these problems. This is the reason for the improved performance of the switched-capacitor filters, both from the standpoint of the reduced initial tolerance and the improved stability with temperature changes.

Very useful adaptations of this basic switched-capacitor integrator have been used to provide a frequency mixing function, a signal summation-mixing-and-filter function, a balanced modulator, rectifier, and even a programmable gain stage.

The availability of analog switches can also provide additional circuit benefits, as shown in Figure 5-7. In Figure 5-7a, an additional analog switch is used to change the polarity of the input signal and to thereby create a noninverting integrator. This is useful because it will now be easy to eliminate the sign-changing amplifier that has been required with the standard biquad RC active filter (this elimination of one of the three op amps used in the resonator loop is a significant accomplishment). A unique new functional possibility is also made relatively simple with analog switches as shown in Figure 5-7b: a *differential integrator*.

a) REALIZING A NON-INVERTING INTEGRATOR

b) REALIZING A DIFFERENTIAL INTEGRATOR

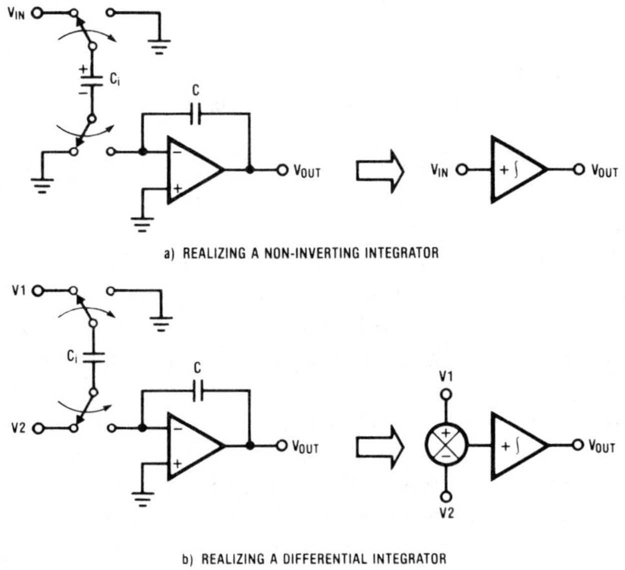

Fig. 5-7 Some Additional Benefits with Switches

Both of these circuit ideas were used by Nello Sevastopolous when he came up with a very clever circuit topology for a general purpose switched-capacitor filter, Figure 5-8. This was a significant contribution to the marketplace, because switched capacitor filters had previously only been used within complex telecommunications circuits and were not

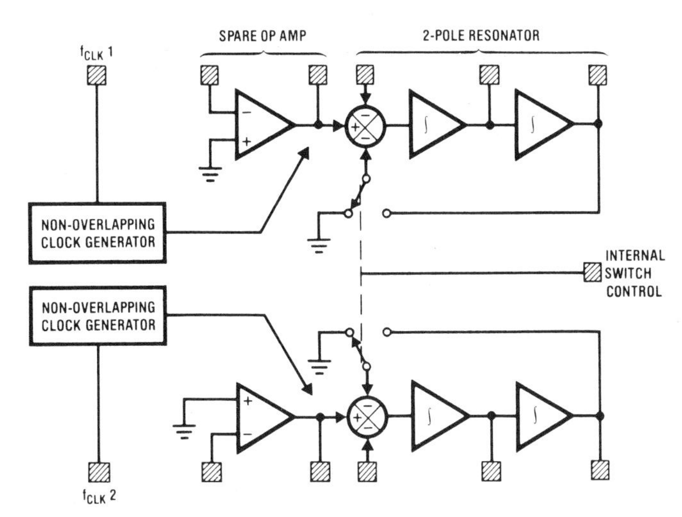

Fig. 5-8 Block Diagram of a General Purpose Switched-Capacitor Filter, the MF-10

available for general filter applications. This first general purpose filter, the MF-10, contained two independent resonators. Each of these resonators could realize a pole pair. They could be used independently, or in cascade to provide a fourth-order filter, using only a single IC package.

Notice that a spare op amp is provided with each resonator. This particular circuit connection easily allows the realization of all of the standard filter functions: low pass, high pass, bandpass, all pass, and band reject. This particular configuration can provide all the application's flexibility that has been realized with a standard three-op amp biquad resonator loop and an extra op amp.

Digital-to-Analog Convertors

The analog switches that are available in a CMOS process allow the realization of low power-drain digital-to-analog converter (DAC) products. The primary function of a DAC is to convert a digital input word into an analog output voltage.

The most popular CMOS DAC circuit approach was first built by Jim Cecil a number of years ago. This circuit did not require the matched base-emitter voltages or current gains that are needed in a bipolar DAC and the low power-drain of the CMOS circuitry helped reduce temperature drifts in the output voltage of the DAC.

An equivalent circuit for these CMOS DACs is shown in Figure 5-9. The resistor network, shown along the top of this figure, is called an *R-2R ladder network* because only R and 2R resistor values are needed to build a DAC of any resolution. The resolution is specified by the number

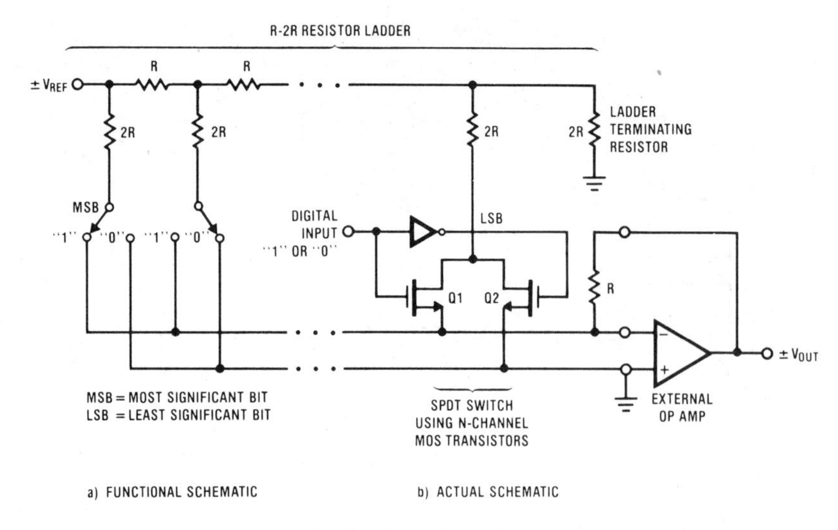

Fig. 5-9 The CMOS MDAC

of digital inputs that are provided: 8-digital inputs provides 256 output voltage steps, 10-digital inputs provides 1024 output voltage steps, etc. Further specifications of a DAC product indicate the accuracy of each of these output voltage steps, and whether or not they always continue to increase or to decrease as the digital input code is incremented or decremented. If the analog output voltage of a DAC always increases (or decreases) as it should, or at worst, stays the same between some adjacent steps, the DAC is said to have a *monotonic* response.

Notice that the single-pole double-throw switches shown on the left side of this figure are realized by the two N-channel transistors shown on the right side of the figure. The conduction state of these transistors is controlled by the digital input signal. The logic inverter insures that only one of these transistors will be on at any given time.

The function of the conducting MOS transistor is to produce a low resistance from source to drain. The direction of the current flow is not important in an MOS transistor. The functions of the source and drain simply interchange for a current reversal. If the resistors used in the R-2R ladder are thin film and therefore do not have a parasitic PN junction associated with their construction, then this DAC can operate with either polarity of V_{REF}. This is the only monolithic DAC to provide this reference polarity freedom. We will now look a little more closely at the operation of this circuit.

A DAC is a mixed-mode multiplier. It provides an output analog voltage that is the *product* of a *digital input word* and an *analog input voltage*. This CMOS DAC can provide a multiplier function that can be operated in all four quadrants, that is, properly accounting for the algebraic sign of both of the inputs, if the digital input is also a signed number. As this is the only monolithic four-quadrant multiplying DAC, it is often referred to as an MDAC or *Multiplying DAC*.

To illustrate the operational simplicity of this DAC, we have redrawn a 4-bit R-2R ladder network separately in Figure 5-10, and have also indicated on this figure the voltages that exist at each node and the resulting current magnitudes that will be provided in each of the 2R legs of the ladder. This analysis can be easily done by proceeding from the right-hand side of this figure—where 2R in parallel with 2R provides the equivalent resistance of R—to the V_{REF} end.

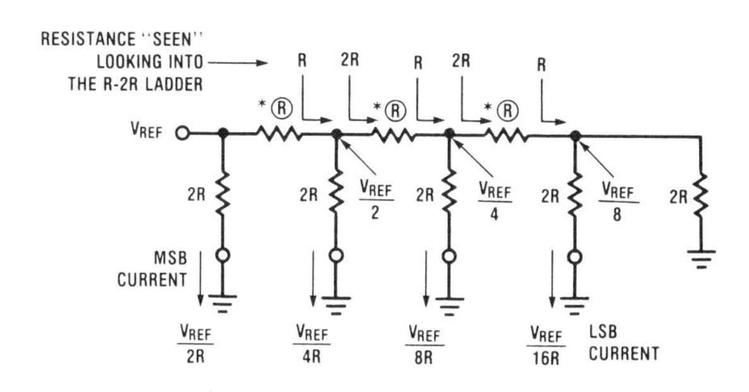

Fig. 5-10 The Binary Current Reduction of a 4-Bit R-2R Ladder

This relatively simple network is seen to provide currents that are successively reduced by a factor of two as we move from the left, or *most significant* side, to the right, or the *least significant* side, of the ladder. The digital input code of the DAC determines whether these binary-weighted currents will be simply shunted to analog ground for a "0" logical input or passed on to the summing junction of the op amp for a "1" logical input.

The op amp converts these binary-weighted currents into an output voltage. The details of this conversion are shown in Figure 5-11 where we are considering the case where only the MSB bit is in the logical "1" state. Notice that this MSB current is the only current that is fed to the summing junction of the op amp; all of the rest of the switches are dumping the other binary-weighted currents to analog ground.

This MSB current then flows through the resistor, R, called *gain* setting resistor, that is serving as the feedback resistor for the op amp (connecting the output back to the inverting input). The flow of this current through this feedback resistor causes a voltage drop, that is shown on the figure, across this resistor. The summing junction (or virtual ground) of the op amp remains as essentially ground ($0V_{DC}$), so an output voltage

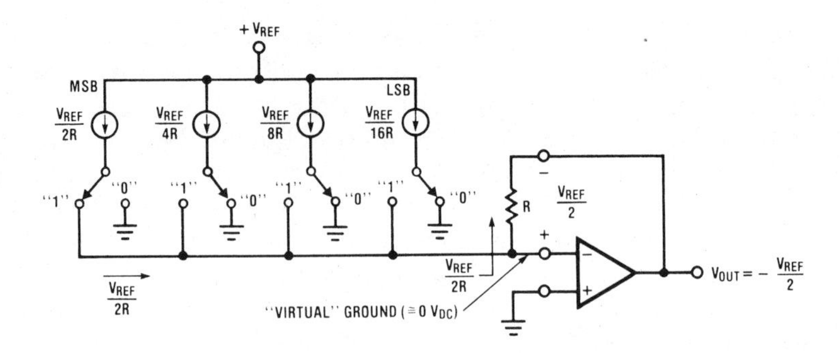

Fig. 5-11 Converting Currents into an Output Voltage

is produced that is $-V_{REF}/2$ as shown in this figure. This is one-half full scale and is the proper output voltage for the MSB acting alone.

Note that the presence of the virtual ground insures that these binary-weighted currents can be individually summed with very little interaction. It is this current summation feature that allows 16 uniformly-increasing output voltage-levels to result from the 4 binary-weighted currents that are shown, if the resistors are precise.

Analog-to-Digital Converters

The analog-to-digital (A/D) converter provides a digital output code in response to an analog input voltage and has been a more difficult circuit to design than the DAC. A/Ds are used to allow analog input voltages to be entered into a digital computer (or a digital controller) in a digitized, computer-readable form. These analog voltages may result, for example, from a temperature sensor. An A/D may have an internal DAC. In addition, a high-performance voltage comparator is needed. The difficulties in the design of this voltage comparator have kept the medium-speed A/Ds behind DACs in bit capability; and, because of an also increased digital complexity, has resulted in higher costs for an A/D. High-resolution low-speed A/Ds, like the "dual-slope" A/D products, are available at relatively low cost.

To achieve a high-speed A/D converter, the comparator must respond rapidly, even with a relatively small input voltage difference (<1 LSB). Uncertainty or oscillations are not allowed in the response of the comparator. Hysteresis cannot be added to the comparator because this would make the performance of the A/D different for an increasing, versus a decreasing, analog input voltage, which is undesirable. Finally, the noise of the comparator must be kept low to reduce the uncertainty of the digital output-code transitions. This combination of performance requirements poses a difficult comparator design problem.

Many unexpected benefits of the sampled-data comparator have solved these performance problems, and also have provided increased functionality in A/D converters. Customer conveniences, such as a differential analog voltage input, a configurable analog multiplexer (Mux), and a reference voltage option of V_{REF} or $V_{REF}/2$ are easily provided. The sampled-data comparator also allows multiple usage of a single resistive-ladder network which greatly reduces chip size. (We will consider this later in this section.)

A basic sampled data comparator (Figure 5-12) consists of CMOS analog switches, a string of capacitively coupled logic inverters for *voltage gain* (only one is shown), and capacitors to convert from voltage to charge and to couple the inverters.

Figure 5-12a shows the node voltages and switch conditions for the *zeroing cycle*, the first part of a sampled-data voltage comparison. The switch that shorts out the gain block, a logic inverter, forces this inverter to operate with $V_{in} = V_{out}$. This is often called the *trigger voltage* (shown as V_B on the figure) of an inverter and serves as a bias reference for this sampled-data comparator.

During this zeroing cycle, the capacitors are charged. The input capacitor, C1, charges to V_B with the polarity shown. The interstage coupling capacitors, such as C2, charge to the small difference voltages that may exist between the V_B values of successive stages.

The *compare cycle*, shown in Figure 5-12b, is entered when all switches change state. All of the gain blocks (the logic inverters) are now fully active, each with a voltage gain of approximately -30, and will amplify any difference from the voltage V_B that exists at the inputs (such as the $+10$ mV that is shown).

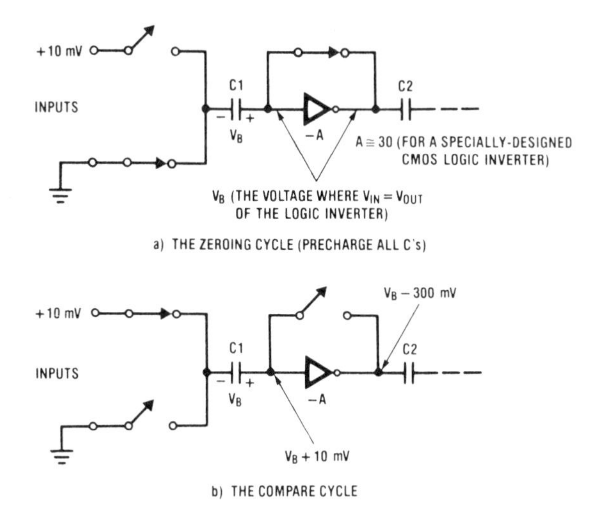

Fig. 5-12 A Basic Sampled-Data Comparator

Notice that the V_B voltage that was previously stored on C1 now adds to the 10 mV input signal. This causes the input voltage directly at the gain block to be 10 mV higher than V_B. The gain of −30 of this inverter therefore provides an output voltage which is 300 mV below the V_B reference. This amplified shift from the V_B values propagates down the cascade of logic inverters. A large enough number of gain stages is cascaded so that the last inverter will provide a full logic voltage level (either 0V or 5V) in response to a desired small input difference voltage.

This basic circuit has been expanded to simultaneously compare multiple differential input voltages, as shown in Figure 5-13. Note that the capacitors C1 and C2 are used to store the values of *one* of each of the *differential* input voltages during this zeroing cycle.

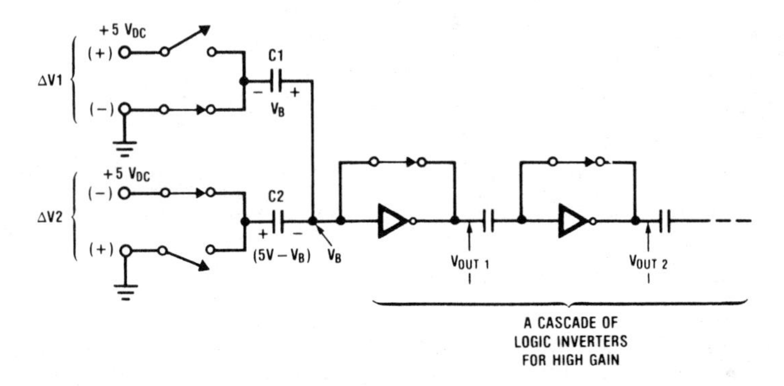

Fig. 5-13 The Sampled-Data Comparator in the Zeroing Cycle

When we switch to the comparison cycle, we have the conditions shown in Figure 5-14. The voltages shown on the capacitors C1 and C2 are the initial voltages. This is the special case where the charges are balanced. The displacement charge (ΔQ1) entering the charge summing node is just balanced by the displacement charge (ΔQ2) that is leaving, so the biasing voltage returns to V_B. Notice that these input charges can be weighted simply by choosing the relative values of the capacitors C1 and C2. (We will make use of this shortly.)

If the upper differential input voltage (ΔV1) had been slightly larger than the lower differential input voltage (ΔV2), then the resulting input voltage to the first gain block, the first logic inverter, at the end of the compare cycle would be larger than V_B. This small difference voltage increase at the input to the first of the logic inverters would then propagate to the last inverter in the string where a full logic voltage level would result. This becomes the digital response to the multiple differential input voltage comparison that was just made.

This novel sampled-data comparator has many application advantages. It has allowed the development of low cost, small IC chip, CMOS A/D

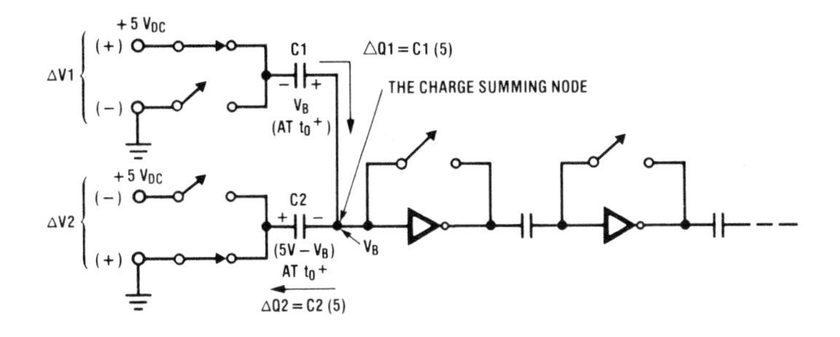

Fig. 5-14 The Sampled-Data Comparator in the Compare Cycle

converters and has provided a differential input-voltage feature that has been used to simplify the internal DAC used in these A/D products.

The differential operation of the sampled-data comparator allows us to borrow the old, low-cost engraving trick that is used in the manufacture of drafting scales, Figure 5-15. Most of this scale has only the major divisions engraved. Only one section is subdivided with the higher-resolution engravings. This reduces the engraving costs, but requires a *differential measurement* to be made. An unknown length is therefore measured by making use of the particular major division that allows the *surplus* of the unknown length to *fall into the higher resolution section*. This is the basic idea of the differential DAC (DDAC), where we save resistors (and their associated die area) and get the same overall resolution, but at lower cost.

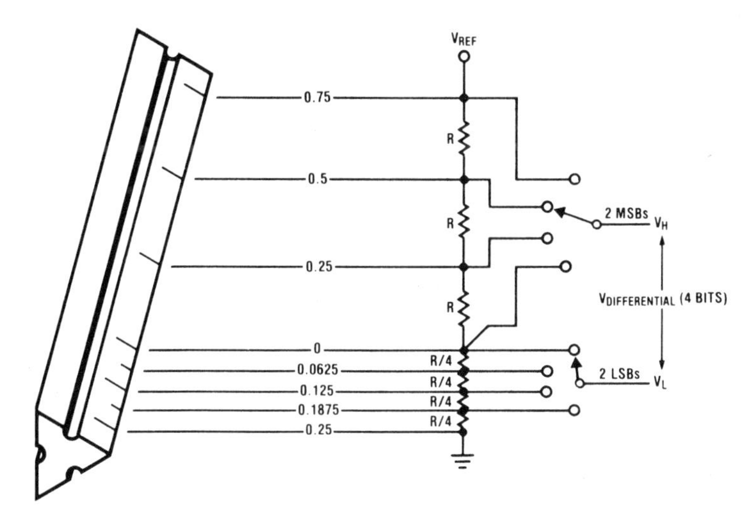

Fig. 5-15 The Differential DAC (DDAC) Concept

Notice that the differential voltage supplied by the decoding switch positions shown on this figure is $0.5V_{REF}$ − zero. This is the first (the MSB) decision in a successive approximation or *binary search* A/D conversion.

The 7-resistor, 4-bit ladder shown in the last figure is replacing a more conventional 16-resistor ladder. The two most significant bits (MSBs) are provided by the upper switch. The next two bits are provided by the lower switch.

We can make use of another cost saving idea with this new sampled-data comparator. It is similar to the sharing of the connecting-rod journals on a crankshaft; the key feature that reduced the overall length of the V-8 automobile engine. In a similar way, here the resistor ladder is shared and we make use of all of the resistor string tap voltages a second time to provide four additional bits of resolution, as shown in Figure 5-16. The proper reduction in significance of these last 4-bits is achieved by scaling down the value of the input capacitor for this second 4-bit group by a factor of 16. The 4-MSBs (shown on the upper right side of the DDAC) therefore use the same capacitor value, 16C, as is used for the analog input voltage. The next 4-bits, shown on the upper left, use an input capacitor value of only C. Now we have an 8-bit A/D and have used only seven resistors and 24 switches for the DAC.

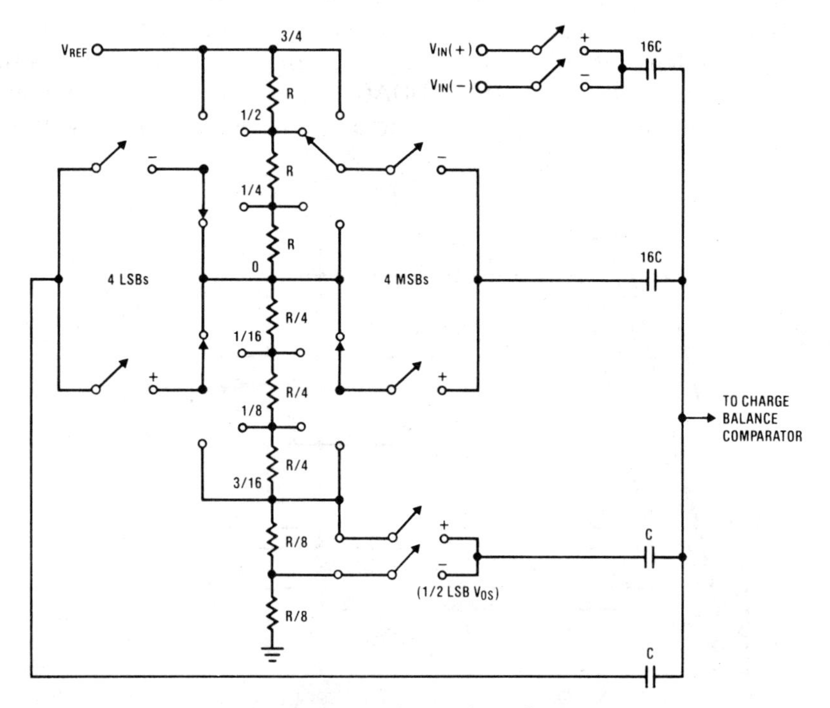

Fig. 5-16 Expanding to 8-Bits by Scaling the Input Capacitors

In this figure, the upper 2-bit switch of the 4-MSB group is selecting 1/2, and the lower 2-bit switch of this group is selecting zero. Again, this provides a differential input voltage for the 4-MSBs of ½ V_{REF}. Notice that zero is sampled by both switches of the 4-LSB group, and this difference voltage is zero. This 4-LSB group will therefore input zero net charge to the comparator.

The extra capacitor and the switches shown in the lower right hand section of this figure always add ½ LSB to the analog input voltage to properly center the ±1/2 LSB *quantization uncertainty* of the A/D. This uncertainty results because a continuous analog input voltage is quantized by the A/D converter. As the analog input voltage is increased from zero volts, this additional input of ½ LSB causes the resulting digital output code to change from all zeros to the digital code of 0000 0001 (1 LSB) when the analog input voltage is only at the value of ½ LSB. In this 8-bit example, 1 analog LSB = $V_{REF}/2^8$ = 5V/256 = 19.53 mV. Continuous analog input voltages over the range from ½ LSB to 1 ½ LSB are *quantized* and all reported as the same digital output code: 0000 0001. This, therefore, has properly centered the ± 1/2 LSB quantization uncertainty that is inherent in A/D conversion about the analog input value of 1 LSB.

An analog multiplexer (Mux) can be easily added to this A/D circuitry by adding extra switches, all of which are connected to the analog input capacitor, as shown in Figure 5-17. Logic circuitry controls which switches are cycled, and thereby selects the analog channel which will be converted. Only two of the switches will be selected and the relative timing sequence of these switches controls the algebraic sign of the differential analog input voltage that results. This allows differential inputs of either sign, for example, channel-0 minus channel-1, or vice versa; a pseudo-differential input voltage feature, such as channel-0 minus COM-mon; or a single-ended channel, as channel-0 minus Analog Ground. The configuration of this *Versa-Mux* (Versatile Mux) is made when a microprocessor writes a *control word* to the A/D at the time the conversion is started. Software control of this Mux allows complete flexibility. Products with 2, 4, and 8-analog input channels are available that make use of this Versa-Mux concept.

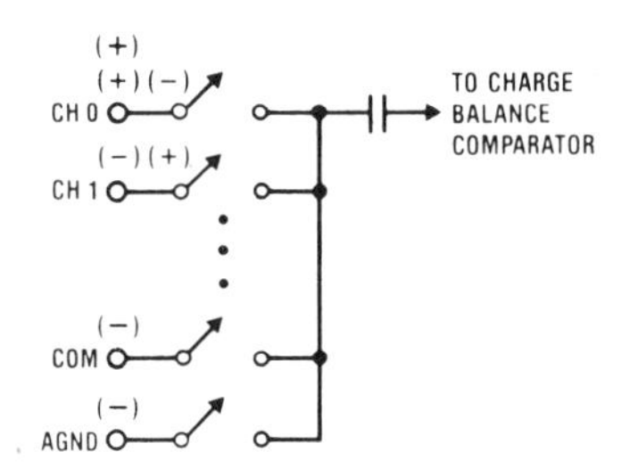

Fig. 5-17 The Versa-Mux added to an A/D

The CMOS sampled-data comparator has been a major contribution to the new A/D products. These converters are now the most popular and are supplied in the highest volume to the industry. The low cost of CMOS A/Ds and the increasing prevalence of digital systems has created a revolution in the application of A/Ds in modern systems.

5.2 A HIGH PERFORMANCE CMOS LOGIC FAMILY

CMOS logic circuits have been available for many years, but the earlier products made use of a relatively old metal-gate CMOS process. Although major strides were being made in the silicon processing that was used for the advanced NMOS products, these improvements were not immediately transferred to the CMOS production lines. In the past, this has caused many people to think of CMOS logic as being limited to slow speeds and as not being competitive with NMOS products.

CMOS wafer processing lines now make use of all of the processing improvements that were initially developed in support of NMOS dynamic RAM products. This has allowed the introduction of a line of CMOS logic circuits (54/74HC) which offer a performance equivalent to bipolar low-power Schottky logic (54/74LS). This 74HC CMOS logic family was considered the *most significant development in logic since the introduction of TTL.*

The basic idea of 74HC is to retain the major advantages of CMOS while providing the key benefits of 74LS bipolar products. Therefore, the higher noise immunity, the wider power supply voltage range, the wider operating temperature range, and the lower power drain benefits of CMOS have all been retained. In addition, 74LS equivalent speeds, pinouts, and output drive current, 4 mA sink and source, have also been designed into this logic family.

There are many system advantages that result from the use of these 74HC logic products: less expensive and smaller power supplies can be used, fans and cooling systems can be eliminated, higher packing density can be used on printed circuit boards which produces physically smaller and therefore lower cost systems, the electronics can be placed in a sealed enclosure, if desired, and the reliability of the circuits is improved as a result of the lower junction temperatures.

The major benefit of CMOS logic is low power drain. This results because power is only consumed when CMOS logic circuits are switched. This power-drain-while-switching results from alternatively charging and discharging stray on-chip capacitances and also from the totem-pole spike current that results from $+V_{CC}$ to ground during the rise and fall times of the digital voltage waveforms.

The Process Used

The 74HC logic family is produced with an oxide isolated, self-aligned silicon gate process. The interconnection problem is not severe with these SSI and MSI products, so only one layer of poly and one layer of metal is used. In addition, many of the layout spacings are enlarged in the mask designs to increase the yield of good circuits.

The Performance Obtained

To appreciate the reason for the 10:1 speed improvement of 74HC, the cross-section of an older metal gate layout for a CMOS logic inverter is shown in Figure 5-18. In this figure, we have also indicated the locations of the important speed-reducing parasitic capacitances that are associated with the circuit. Note the extra P^+ diffusions that are used around the N-channel device, and the extra N^+ diffusions that are used around the P-channel device. These are the *guard rings* (channel stops) that are necessary to prevent the possibility of parasitic MOS transistors in the layout.

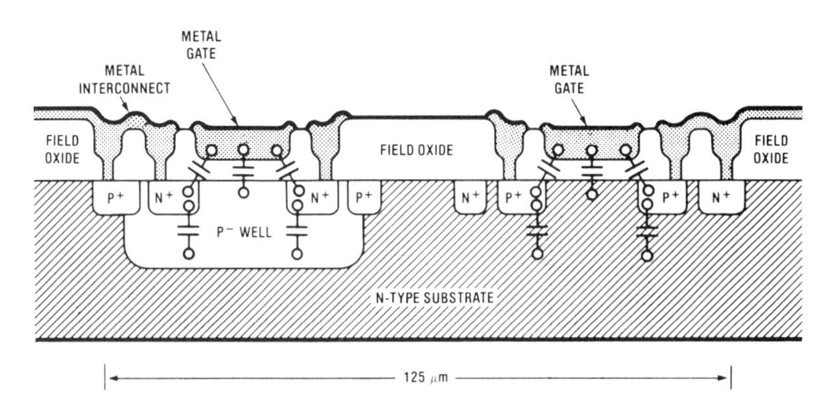

Fig. 5-18 A Cross-Section of a 74C Metal-Gate CMOS Logic Inverter

In this figure, there are separations shown between adjacent guard rings and also between these guard rings and the sources and drains of the MOS transistors. This is to allow room for the depletion layer spreads that result when operating with high-voltage power supplies. For low-voltage CMOS products, these guard rings are *butted*. They are allowed to touch each other and also to touch the source and drain regions of the transistors in order to reduce the size of the logic chips. The horizontal length indicated on this figure for one logic inverter is 125 μm.

A cross-section of the newer 74HC CMOS logic inverter is shown in Figure 5-19 (note that these figures are not drawn using the same scale

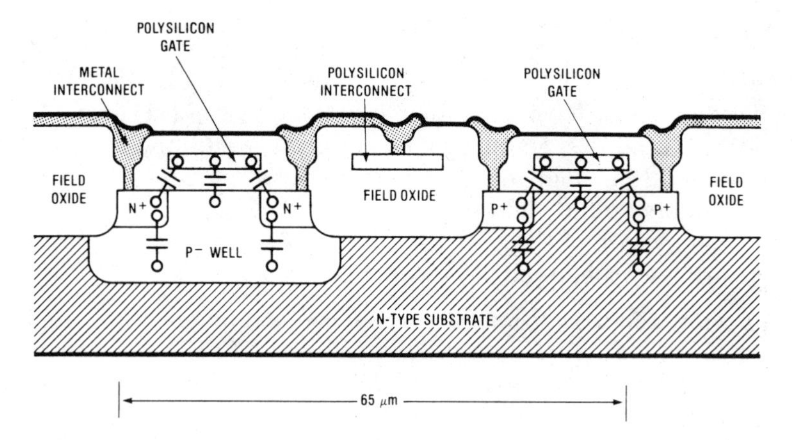

Fig. 5-19 A Cross-Section of a 74HC Silicon-Gate CMOS Logic Inverter

because the factor of two reduction would make this figure illegible). Notice that the guard rings are not used and a field oxide layer is, instead, butted against the sources and drains. This relatively thick oxide layer also prevents the possibility of parasitic MOS transistors in the chip layout. This change to *oxide isolation*, plus the smaller feature sizes that are used, allows the 74HC inverter to occupy a horizontal length of 65 μm, or about one-half the length of the older inverter. This approximate 2:1 reduction of both dimensions reduces the circuit area by approximately 4:1.

This combination of oxide isolation, shallower junction depths, and smaller feature sizes reduces all of the stray parasitic capacitances of a 74HC inverter by at least a factor of two. A comparison of these capacitance values for the two CMOS inverter layouts is shown in Figure 5-20.

The self-aligning silicon gate also reduces the gate overlap capacitances. All of these reduced capacitances, and the approximately 4:1

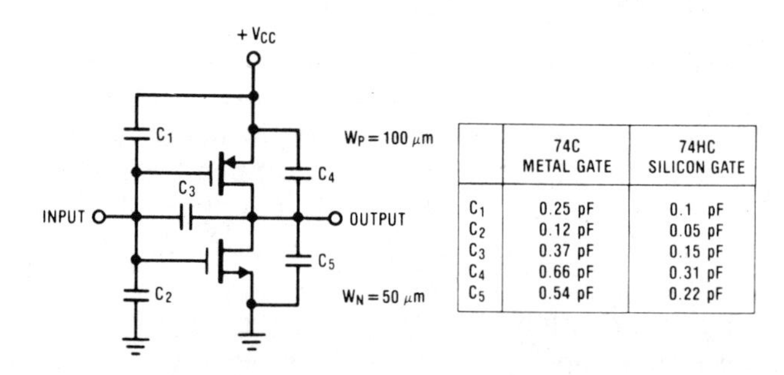

	74C METAL GATE	74HC SILICON GATE
C_1	0.25 pF	0.1 pF
C_2	0.12 pF	0.05 pF
C_3	0.37 pF	0.15 pF
C_4	0.66 pF	0.31 pF
C_5	0.54 pF	0.22 pF

Fig. 5-20 Comparing the Stray Capacitances of a Metal-Gate and a Silicon-Gate Logic Inverter

increase in transconductance that results from the shorter channel lengths and the thinner gate oxides of the 74HC CMOS transistors provides the speed improvement of eight to ten times over that of the metal gate CMOS products.

The output transistors of the 74HC logic family have been designed to meet or exceed the output current that is supplied by the 74LS products. A comparison of output sink current is shown in Figure 5-21, and output

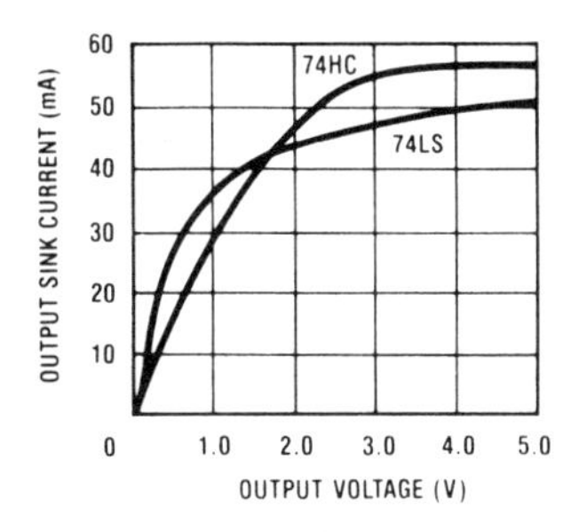

Fig. 5-21 Comparison of Output Sink Current

source current is compared in Figure 5-22. This matching of the output current capability of 74LS provides 74HC with the same ability to drive external load capacitances.

The comparison of the variation of the propagation delay as the load capacitance is increased is shown in Figure 5-23. Notice that the 74HC

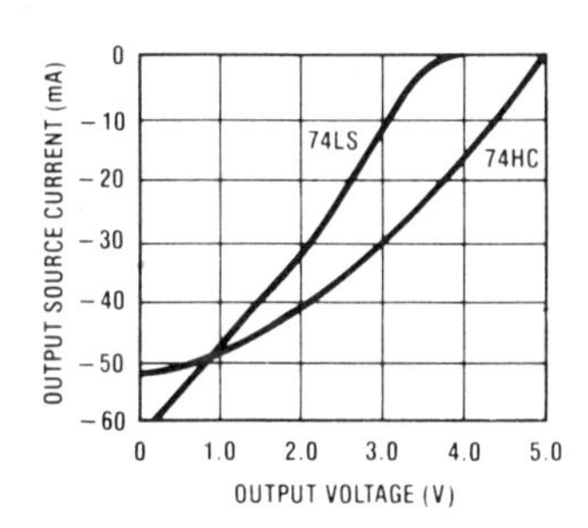

Fig. 5-22 Comparison of Output Source Current

CMOS logic circuit is handling load capacitance as well as, if not better than, the 74LS product.

A comparison of propagation delay changes with ambient temperature variation is shown in Figure 5-24. The propagation delays from *high to low* (t_{phl}) and *low to high* (t_{plh}) are different for bipolar logic (DM74LS00), but are the same for both the metal gate (MM74CXX) and the MM74HCXX CMOS products. Both the metal gate and the 74HC silicon gate CMOS

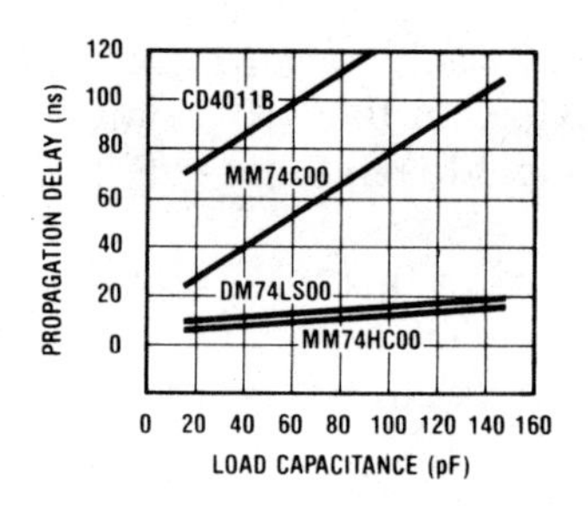

Fig. 5-23 Propagation Delay Dependence on Load Capacitance for CMOS 2-Input NAND Gates

logic products increase propagation delays at approximately 0.3%/°C. (A check of these curves will verify this, although it may not appear so at first glance.)

All CMOS logic circuits suffer the same decrease in speed as the power supply voltage is reduced. This can be approximated as:

$$t_{pd}(V_{CC}) = t_{pd}(+5\ V_{DC}) + 3.5 \times 10^{-1}\ t_{pd}(+5\ V_{DC})[5 - V_{CC}]$$

where:

$t_{pd}\ (V_{CC})$ is the propagation delay at a particular power supply voltage, V_{CC}

$t_{pd}\ (+5V_{DC})$ is the propagation delay measured with $V_{CC} = +5V_{DC}$

This amounts to a propagation delay increase by 3.5% per 100 mV of decrease in power supply voltage.

The larger output currents of the 74HC CMOS logic family also require the same bypass capacitors across the power supply lines that would be used for a 74LS logic board. Also, the RFI (*radio frequency interference*) of 74HC compares to the bipolar logic (the older metal gate CMOS was quieter because of the slower response times). Finally, notice that unin-

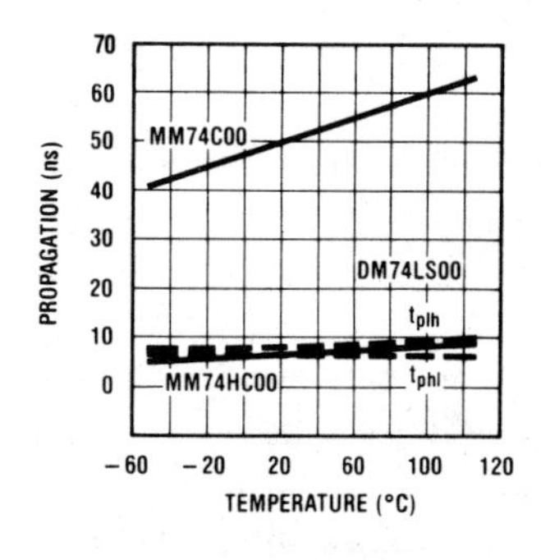

Fig. 5-24 Propagation Delay Dependence on Ambient Temperature

tentional output short circuits to either the power supply line or to ground will result in much larger output currents than those that were obtained with metal gate products.

A special interface must be used from TTL logic when interfacing logic from TTL to the inputs of this CMOS logic family. One of the special TTL to CMOS buffers that are supplied in this family can be used, the 74HCT products. At the output, a fan-out of ten is provided for the interface from 74HC CMOS logic to TTL (74LS) inputs.

Electrostatic Discharge Improvement

Electrostatic discharges (ESDs) result from the charge that people accumulate on their bodies and then dump (with the associated 1- to 2-inch spark) into electronic circuits when they touch a piece of electronic equipment. The voltages involved can be as large as 35 kV, so it is no wonder that ICs have trouble surviving this harsh treatment.

Testing for the IC circuit tolerance level for ESD generally makes use of the circuit shown in Figure 5-25.The 150-pF capacitor is repeatedly charged to a high voltage, and then this charge is sequentially dumped through a 1.5-kΩ series resistor into all input and output pins of the IC package. The maximum value of voltage that can be used without damaging the IC is then noted.

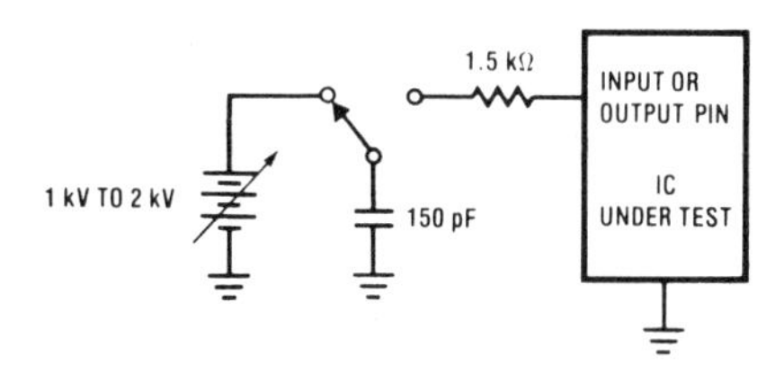

Fig. 5-25 The Standard ESD Test Circuit

ESD can melt silicon, rupture SiO_2 layers, fuse the interconnect aluminum metallization, cause polysilicon to evaporate, and can charge a surface state. These effects most likely will destroy an IC, but low level ESD can just cause changes in leakage currents and offset voltages. These more subtle disturbances are more noticeable with linear ICs.

To improve the ESD immunity (and also the SCR immunity, which we will discuss next), poly resistors are used in series with each input pin, as shown in Figure 5-26. In addition, relatively large geometry diodes are added as clamps on the IC side of these input poly resistors (and also at the output) to restrict the magnitudes of the voltages that are applied to the rest of the IC circuitry. These layout enhancements have allowed CMOS logic to withstand ESD test voltages in excess of 2000 volts.

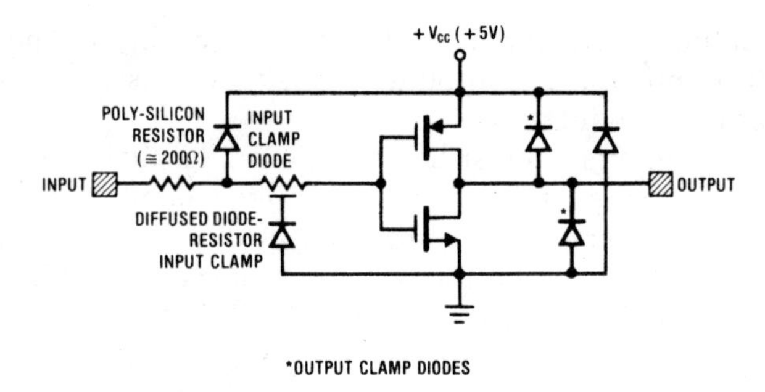

Fig. 5-26 Improving ESD and SCR Immunity

Improving SCR Immunity

The parasitic *Silicon Controlled Rectifier* (SCR) that exists in a CMOS layout involves the sequence of four alternating-types of doped regions that was described in Section 2.4. The use of an epi-substrate in the fabrication of CMOS logic allows a guaranteed spec of ±20 mA of current to be either pushed into or removed from any input or output pin without triggering an SCR. This spec applies over the complete temperature range of these products.

More work is needed in the area of reducing the SCR problem in CMOS products. It is expected that an increased emphasis on layout techniques will be used to supplement (and to possibly eliminate) the use of epi-substrates. This extra care may be needed because SCR problems are increased as the channel length drops to 1 μm and below (because the smaller spacings increase the current gain, β, *of the parasitic bipolar transistors that form the latch circuit*).

5.3 SPECIAL CMOS CIRCUITS FOR THE NEW TELEPHONE SYSTEMS

The telephone system has evolved as an extremely large installed base of wiring networks and special buildings to house the required electric and electronic equipment. As the demand for telephone service continues to rise, ways must be found to increase service in the most economical manner. The first thought is to find a way to use the existing wiring networks and buildings and to somehow add to the number of subscribers that can be placed on this existing network. Thus, a time-division multiplexing, pulse code modulation (PCM) digitized-voice scheme is being implemented where more than 24 subscribers can be simultaneously placed on a single wire-pair that used to accommodate only two subscribers in the previous analog system. This move to the modern digitized telephone system has created a large demand for specialized ICs to digitize the voltage waveforms that are derived from human speech.

The telecommunications industry is a large growth area for ICs because the major changes in the telephone system require specialized, low-cost, complex ICs. This old, traditionally analog system is finally being converted to digital.

The conversion to the complete digital telephone system is being done in stages. The *trunk lines*, which interconnect the telephone exchanges, were first converted to PCM. In the future, each subscriber's individual telephone will contain added circuitry to digitize the voice signal prior to transmission to the telephone exchange. Additionally, circuitry will also be provided within each telephone to convert a received digital bit stream back into a voice voltage waveform to operate the earpiece in the telephone handset.

Benefits of Digitized-Voice Transmission

Digitized voice transmission has advantages over analog voltage voice transmission because digital waveforms can be passed through a large number of electronic repeaters without introducing distortion. Limits do arise as a result of noise; but, unlike an analog voltage line-amplifier, each digital repeating amplifier is more likely to pass on the same bit stream that it receives.

There are many ways to use digital techniques or pulses to communicate the measure of a voltage amplitude. In many of these schemes, one characteristic of the transmitted pulse is modulated or varied by the magnitude of the voltage information; such as the *amplitude* (pulse amplitude modulation, PAM), the *width* (pulse width modulation, PWM), or the *position* in time (pulse position modulation, PPM). All of these schemes are referred to as being *uncoded* because only one pulse is used to convey the transmitted information. Therefore, any small deformations in the received amplitude, width, or position, respectively, of these pulse modulation techniques can introduce noise into the received signal.

If several standard pulses are, instead, transmitted for each voltage amplitude sample, a reduction in noise contamination can be exchanged for the increase in the transmission bandwidth that is required for these extra pulses. This is the benefit of a *coded* scheme, such as pulse code modulation, PCM. In fact, in PCM the operations of periodically sampling the voice waveform, approximating each sample to the nearest permitted voltage reference level (quantizing), and encoding the sample as a series of pulses are important in determining the quality of the communication link.

The largest problem in PCM is the process of approximating or quantizing the sampled amplitude value. This results in *quantization noise*. Increasing the total number of amplitude levels that are used (increasing from 8 bits, 256 levels, to 10 bits, 1024 levels, for example) reduces this quantization noise at the expense of the increased bandwidth that is necessary to transmit the extra coding pulses. In PCM, this signal-to-quantization-noise power ratio increases exponentially with bandwidth—which is an extremely efficient exchange that approaches the theoretical maximum that can be obtained.

In PCM transmission, this inherent quantization noise is nonaccumulative. This is in contrast to the uncoded pulse modulation techniques. Noisy PCM pulses can be easily regenerated or restandardized and so the path length or the noise that is introduced by the communication link becomes unimportant.

Other benefits become possible once the decision to convert to a digital system is made. For example, the increasing interest to place digital data on the existing telephone lines is easily done when the voice transmission is also implemented with digital bit streams.

Further benefits of a digital telephone system is the simplified switching problem, and the ease of adding additional control features such as complex automatic cost accounting, call holding, call waiting, call forwarding, call transfer on busy, temporary transfer, conferencing, abbreviated dialing, and automatic dialing of stored telephone numbers.

The Telecom Problem

The circuits used in the new telephone systems must be very power conserving. Large numbers of individual circuits are now used, and the complete system must still have auxiliary power sources in case of the temporary loss of the local ac electric power source. The high cost of these back-up power sources (batteries) places a premium on low power-drain electronics.

CMOS processes were selected for telecom products because CMOS can combine linear circuits with the digital circuits that are needed. A very important additional factor is the resulting low power drain of the CMOS circuitry.

Coding and Decoding

The encoding of the human voice into a digital form and the decoding of a digital bit stream back into an analog form is handled by coder-decoder circuits called *codecs*. An analog sample-and-hold circuit captures time samples of the input speech voltage waveform and an A/D converter provides a digitized version of each of these successive speech voltage waveform samples. These digital signals are then sent in a serial format over the telephone network.

Increased resolution for small amplitude signals is provided by a nonlinear encoding technique so that only 8 bits are needed for each time sample. Human speech contains a high density of low voltage samples and nonlinear encoding more nearly equalizes the probability of obtaining any code. This improves the information transfer.

The audio input signal is sampled at an 8 kHz rate and special time-slot assigning circuitry time-division multiplexes many of these digital signals over a single pair of wires. A DAC is used to convert from the received digital information to an analog voltage for the earpiece in the telephone handset.

Both the audio input signal to the A/D and the recovered audio signal from the DAC are passed through special filter circuits. These filters make use of high-order switched-capacitor circuits.

The codec and the filters are placed on a single *combo* chip. These combos consume only 50 mW of power from $\pm 5\ V_{DC}$ supplies and can be placed in a power saving 1-mW standby mode.

5.4 CMOS MICROPROCESSORS AND MEMORY PRODUCTS

The most important CMOS attribute for microprocessors, microcontrollers, and memory products is its low power consumption. This is increasing the application areas for these products because many high performance battery-powered computer systems are now possible. Existing microprocessors, such as the Z80™*, have been made available in a CMOS equivalent and advanced CMOS processes are now used for essentially all of the newer microprocessor and memory products.

In those applications where low power drain is more important than the operating speed, the clock frequency can be reduced with these CMOS processors and a significant savings results because the power consumption of CMOS varies directly with the clocking frequency. Additional operating power savings can be obtained by reducing the power supply voltage to less than $5\ V_{DC}$.

CMOS Dynamic RAM (DRAM) has been reported (ISSCC 1988) that has achieved single chip densities of 16 Mbit. This uses an advanced 0.5-μm technology with a high-capacitance trench structure that surrounds the active devices. Many researchers are making use of deep trenches (for isolation, capacitors, and even active devices) to increase the chip packing density by making use of the vertical dimension (down into the silicon). It is interesting that this single chip 16-Mbit DRAM is $93.85\ mm^2$, will fit in a standard 300-mil DIP, and has twice the storage capacity of l-Mbyte floppy disk. Up to 256-Mbit DRAM has been projected for the future using a 0.25-μm process.

*Z80 is a trademark of Zilog Corporation.

5.5 DIE COATINGS FOR IMPROVED PRODUCT RELIABILITY

Plastic packaged ICs are lowest in cost, but this package does not provide the sealed protection from ambient contamination that exists with a hermetic (ceramic) package. Major improvements were made in plastic-packaged products with the introduction of an additional die-surface coating. The theory is that if cracks, pin-holes, or other problems exist in the surface-protecting oxide layer on the chip, an additional surface coating will seal these openings and therefore protect the die.

The plastic packaging problem is made worse by the relatively light impurity doping levels that are used in MOS products. Lightly doped silicon can be easily affected by small concentrations of undesired impurities. An effective solution to this problem (for CMOS logic products) has been to add a final die coating layer that effectively seals the surface and improves the reliability of plastic encapsulated MOS products.

5.6 A MULTIPLE-LAYER-METAL CMOS PROCESS

To indicate the processing complexity of an early CMOS process, we will use the multiple-layer-metal process that was originally developed for CMOS gate arrays.

The older isotropic wet-chemical etches of the oxide layers and aluminum interconnect metal layer on the wafers are replaced by non-isotropic (vertically etching) dry etching techniques. Also, the diffusion furnaces which were used to introduce doping impurities into silicon are replaced by better controlled ion implantion machines. Many special masking techniques are also now used to obtain high-resolution lithography. From 12 to 15 separate masks are used in these processes.

As an example of the processing steps used to obtain a multiple-layer-metal, N-well CMOS product, we show cross-sections of a small region of a wafer and highlight some of the steps that are typically used in wafer fabrication. For simplicity, we will not use an epi wafer, as discussed earlier for SCR protection, as the starting material. The conventional starting wafers are relatively lightly doped (have a resistivity of approximately 8 Ω-cm), and use a ⟨100⟩ crystal orientation.

In Step 1, Figure 5-27, a P-doped silicon wafer has an SiO_2 layer grown on the surface by placing the wafer in an oxidizing furnace where an elevated temperature and a gas ambient that contains oxygen are provided. The wafer surface is then covered with a thin layer of photoresist. This is generally done by spraying a controlled amount of photoresist on the wafer and then spinning the wafer to smooth the resist and throw off the excess. The photoresist is then slightly baked to toughen the film.

In Step 2, Figure 5-28, the photoresist is selectively (by use of a mask) exposed to ultraviolet light. The ultraviolet light passes through the clear

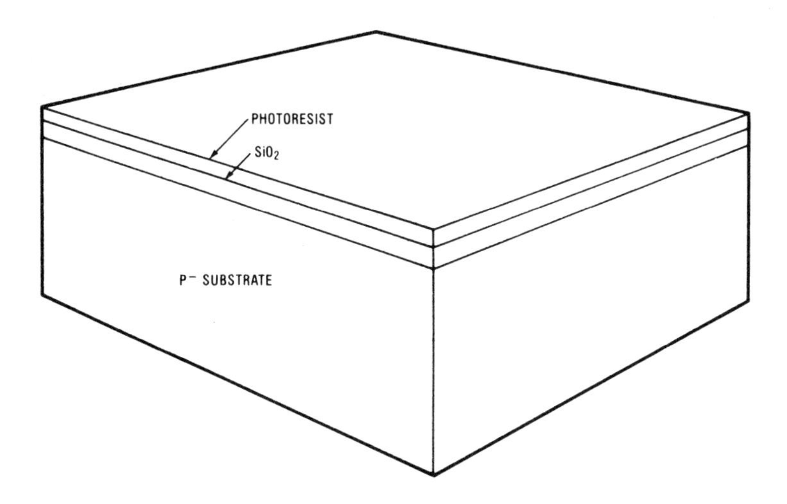

Fig. 5-27 Step 1: Oxidize Surface and Coat with Photoresist

regions of the mask. The opaque (black) regions of the mask block the light in selected small areas across the silicon surface. For clarity, the example here is using *negative resist* and the unexposed resist regions will be dissolved away in an organic solvent: the *developing* operation.

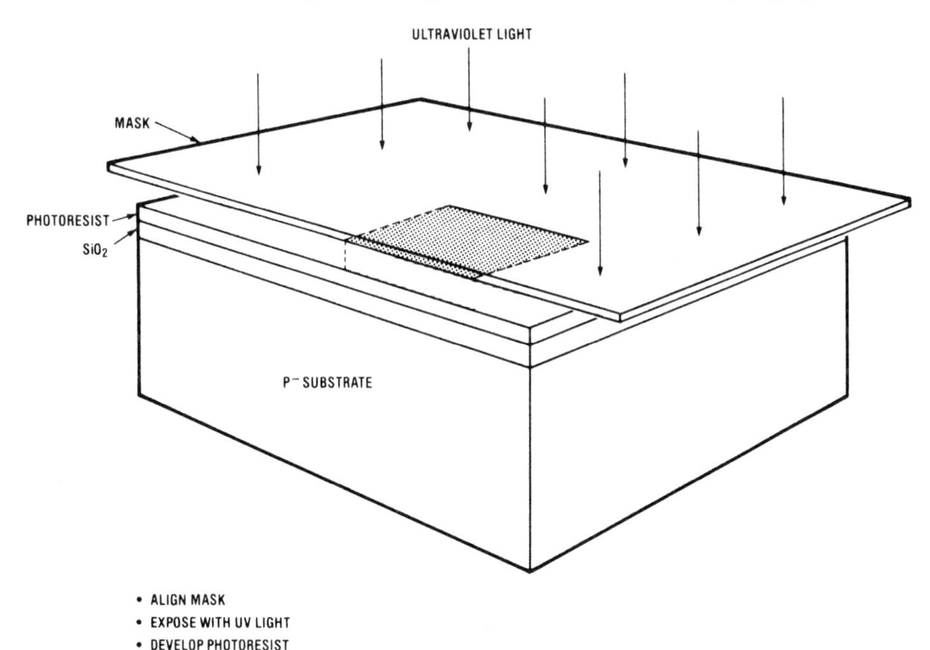

Fig. 5-28 Step 2: Etch Oxide Layer

Positive resists are used for small feature size and the opaque and clear areas of the mask are reversed because exposure to ultraviolet light with positive resists causes removal of the photoresist in the developing operation. In either case, the remaining resist pattern then serves as a protective film (like a varnish) and allows the etching process that follows this step to clear away only the uncovered or exposed regions.

After selected areas of the oxide layer are removed by etching, the silicon wafer is exposed to an electrostatically directed beam of energetic ions: the ion implantation processes, Step 3, Figure 5-29. (The photoresist also remains as an implant mask.) These dopant atoms only enter the silicon where it is exposed to form the N-well regions in which the P-channel transistors will eventually be located.

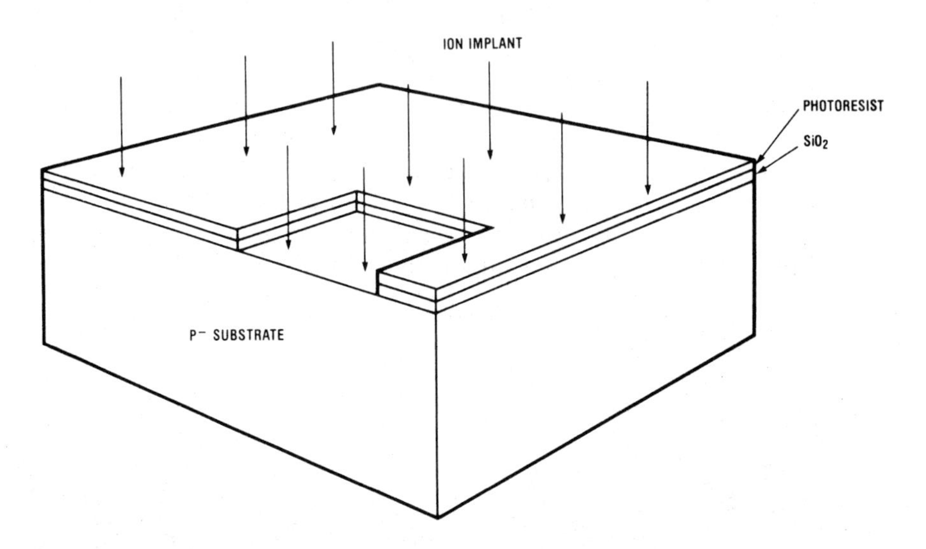

Fig. 5-29 Step 3: Implant for N-Well

After the photoresist layer is removed, a high-temperature cycle causes the implanted atoms to diffuse further into the silicon and to take substitutional positions in the silicon crystal lattice. This actually forms the N-wells, as shown in Step 4, Figure 5-30, and also gives the surface of the silicon crystal a chance to recover from the physical damage that was inflicted by the high-energy implant ion bombardment. An oxide layer is simultaneously formed in the furnace to seal the openings.

In Step 5, Figure 5-31, the oxide is etched away and a new thin oxide is grown. A layer of silicon nitride is then deposited over this thin oxide.

A sequence of processing steps is indicated in Step 6, Figure 5-32, that prepares for the doping of the field regions (the areas between the transistors) of the wafer. The silicon nitride (Si_3N_4) layer and the under-

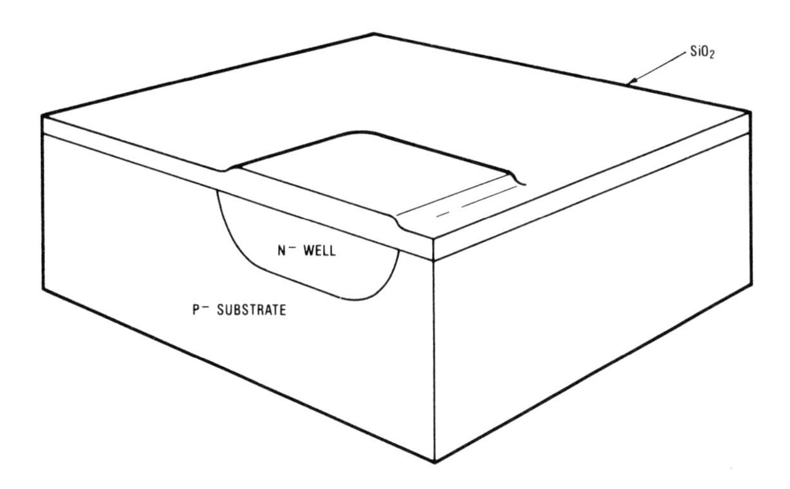

Fig. 5-30 Step 4: Anneal Implant and Grow Oxide

lying oxide layer are used to cover all of the areas in which the transistors will eventually be fabricated. This nitride layer masks the field implant doping (prevents field implanting the transistor areas) and also prevents the growth of field oxide over the transistor areas.

In Step 7, Figure 5-33, the wafer is covered with photoresist and this photoresist is then masked and developed to expose the areas for the next implant. This layer of photoresist, and the previous nitride/oxide sand-

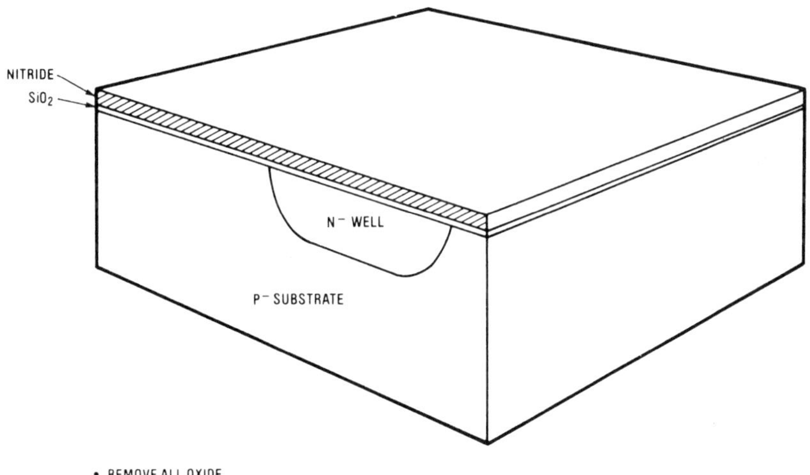

Fig. 5-31 Step 5: Deposit Nitride Layer

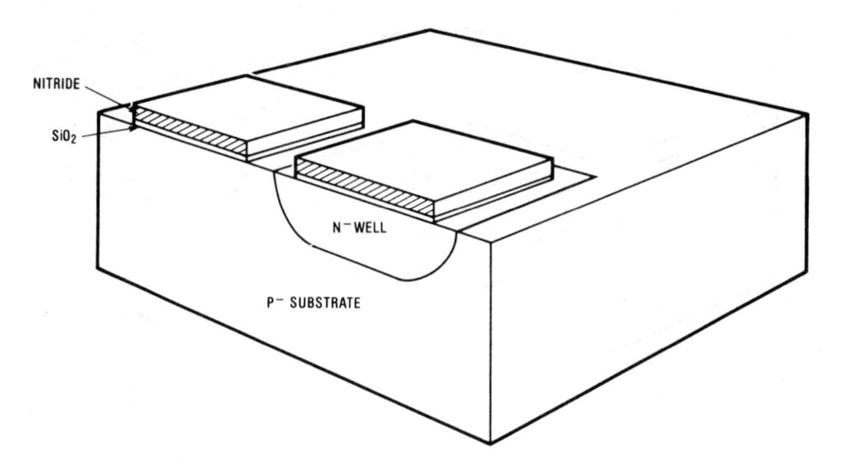

Fig. 5-32 Step 6: Selectively Remove Nitride and Oxide Layers

wich, block the implant ions in both of the transistor regions. The rest of the wafer receives a P-type (boron) field-dopant implant to prevent parasitic MOS transistor action. An N-type of field doping is not necessary in some processes because the N-dopant phosphorus naturally tends to *pile up* at the silicon surface (next to the SiO_2 layer). In contrast to this, the P-dopant boron naturally tends to diffuse into the SiO_2 layer. This loss of surface doping in the P-type regions therefore requires the extra P-type field implant step.

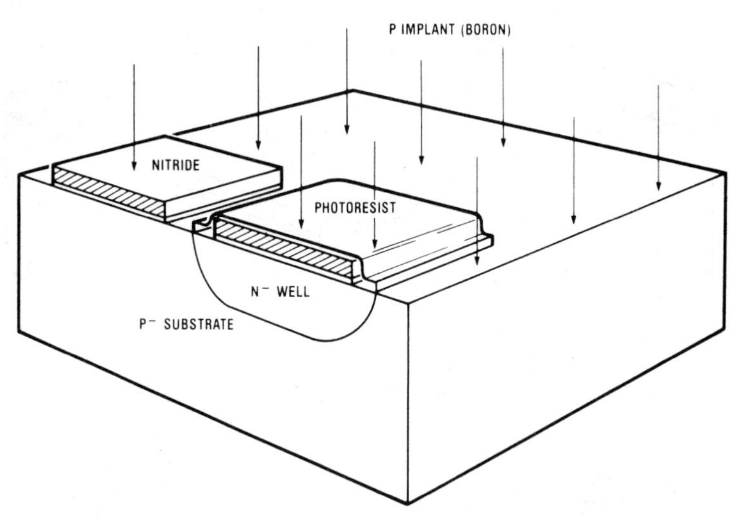

Fig. 5-33 Step 7: Implant the Field Regions

In Step 8, Figure 5-34, following the P-type field implant, the wafer is placed in an oxidizing environment and a thick layer of field oxide is grown in the regions where the nitride has been removed.

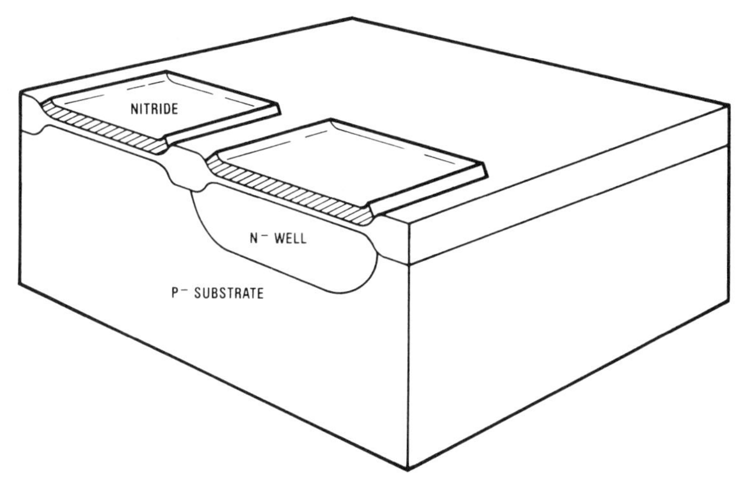

Fig. 5-34 Step 8: Grow Field Oxide

In Step 9, Figure 5-35, the nitride and the underlying thin layer of oxide are selectively removed (thus exposing the silicon surface of the transistor areas) and a very thin gate oxide is grown. This is the critical

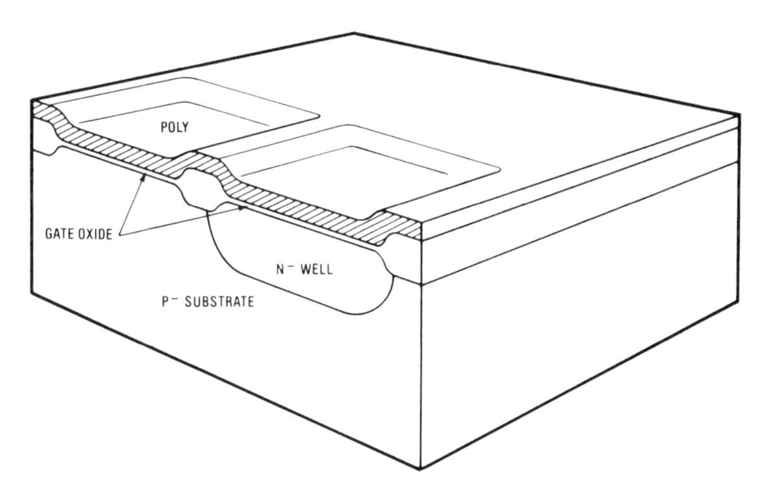

Fig. 5-35 Step 9: Deposit and Dope Poly Layer

step in MOS processing. Ultra-clean tubes are used to thermally grow this low-temperature oxide layer and extreme care is continually exercised to prevent wafer contamination by sodium, particulates, and heavy metals. In addition, this oxide layer must also be very uniform in thickness across the wafer in order to keep the threshold voltages of the MOS transistors uniform. Special low-energy, light doses of ion implanting can now be applied directly through the thin gate oxide layer, using photoresist as a mask, to shift or adjust the threshold voltages of the MOS transistors. The damage from this implant does raise the noise of the implanted MOS transistors, as compared with the "native" (nonimplanted) transistors. This is not a problem for digital circuits, but is a consideration in linear MOS circuits.

The processing for the self-aligned silicon gates now begins. A layer of poly silicon is deposited and then heavily implanted with phosphorus. In Step 10, Figure 5-36, this poly is mostly removed except where the gates are to be located. Photoresist is then applied and a mask is used to define this photoresist and cause it to be removed where N^+ regions are desired, the drains and sources of all of the N-channel transistors. The N-type dopant arsenic is then implanted through the gate oxide using photoresist as the mask over the P-channel transistor regions and the poly

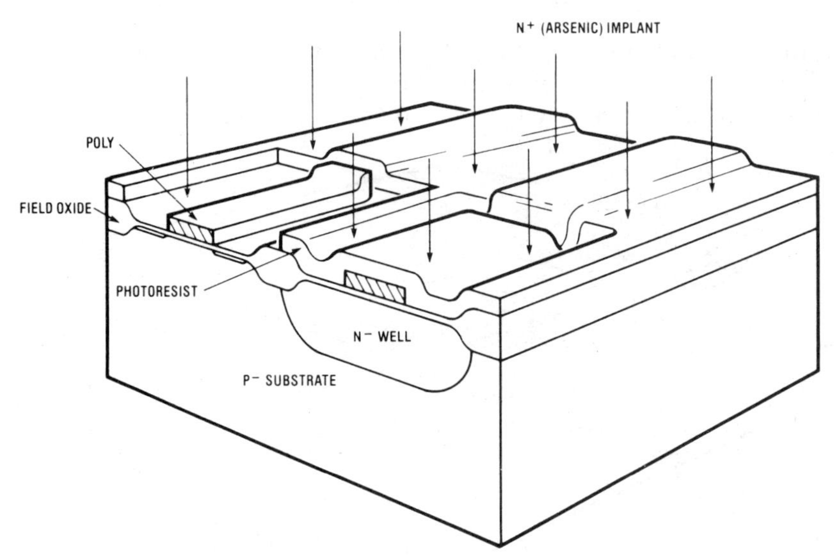

- LEAVE POLY IN GATE REGIONS
- COVER WITH PHOTORESIST
- SELECTIVELY REMOVE PHOTORESIST TO EXPOSE REGIONS FOR N^+ IMPLANT
- IMPLANT ARSENIC
- REMOVE PHOTORESIST

Fig. 5-36 Step 10: Implant Sources and Drains of N-Channel Transistors

gates as masks to define the channel lengths of the N-channel transistors. This photoresist is then removed.

In Step 11, Figure 5-37, using the same sequence of steps as were used in the figure above, the P^+ areas are implanted with boron. This provides the P^+ drains and sources for all the P-channel transistors. The photoresist is then removed. This completes the self-aligned gate processing.

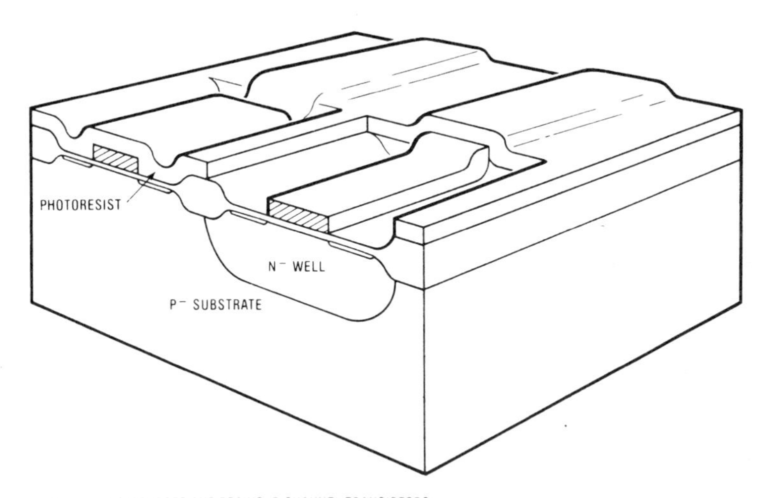

- IMPLANT P^+ SOURCES AND DRAINS (P-CHANNEL TRANSISTORS)
- REMOVE PHOTORESIST FOLLOWING IMPLANT

Fig. 5-37 Step 11: Implant Sources and Drains of P-Channel Transistors

In Step 12, Figure 5-38, a thick oxide layer is deposited over the entire wafer. This glass has a composition of 8% phosphorus by weight so that it will "reflow" at a relatively low temperature (1000°C). An extra "oxide reflow" step then is used to allow this glass layer to smooth or round off all of the previous abruptly etched oxide steps so that there will be no problem with "step coverage" when the metal interconnection layer is applied.

In Step 13, Figure 5-39, the back of the wafer is etched clean to allow contact with the substrate for in-process testing and to remove the material buildup that has accumulated in processing. The wafer is then placed in a high temperature phosphorus (an N-type dopant) atmosphere, a diffusion furnace, which drives in the P^+ and N^+ drain and source regions, dopes the deposited glass, and also dopes the backside of the wafer. This is one of the only uses for a diffusion furnace.

Both the N-channel and the P-channel transistors have now been formed, and all of the transistors now exist under a protective oxide layer that covers all of the source and drain regions. Notice that oxide isolation

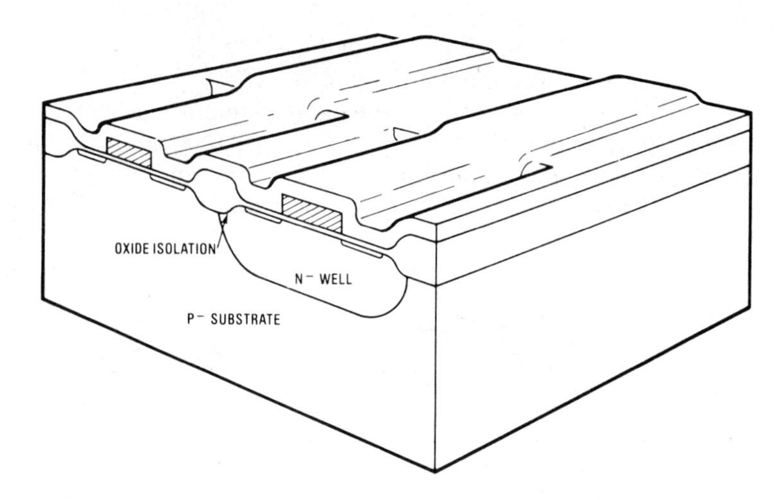

Fig. 5-38 Step 12: Add an Oxide Layer

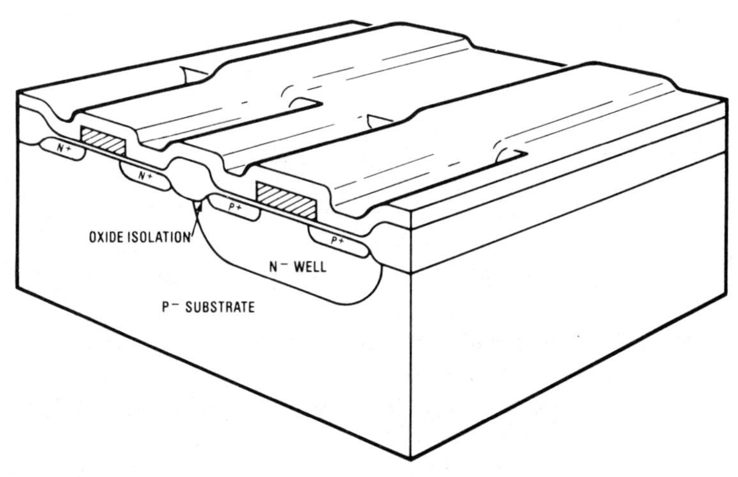

Fig. 5-39 Step 13: Simultaneously Anneal Both Implants

is provided along the edges of all of the source and drain regions of both transistor types. This reduces the parasitic capacitance and therefore improves the speed of the circuits.

In Step 14, Figure 5-40, a contact mask is used to define the areas where the interconnect metal is to contact silicon or poly. The oxide is then etched away from these areas.

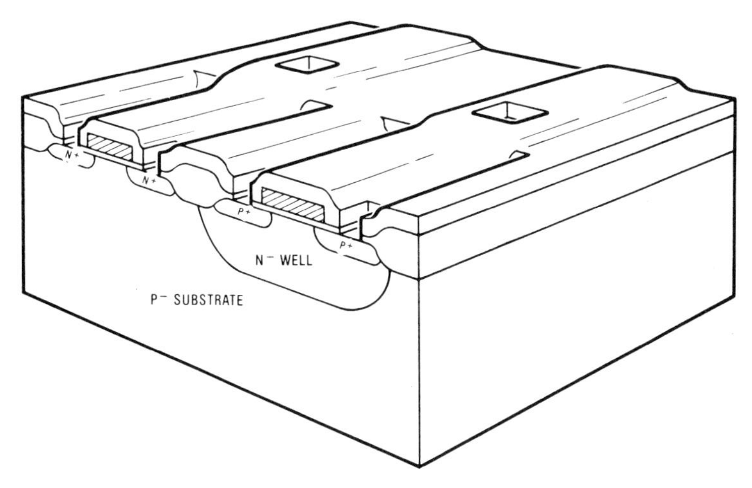

Fig. 5-40 Step 14: Open Contact Windows

In Step 15, Figure 5-41, an aluminum alloy is deposited over the entire wafer. This first metal interconnect layer passes down through the contact holes and provides electrical connection to the sources, drains, and the poly gates. A dry etching step then removes all of the undesired metal and leaves behind the particular Metal I interconnect pattern that is desired for the intended circuit function.

In Step 16, Figure 5-42, a second insulating oxide layer is deposited to cover the first metal pattern. Very small geometry contact windows, or

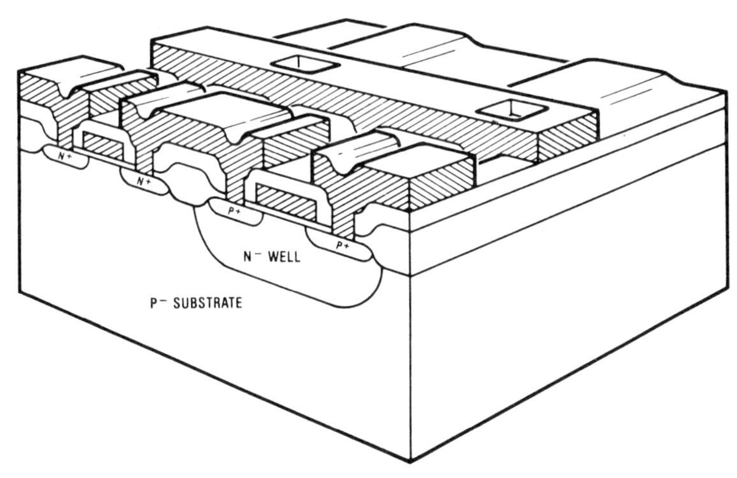

Fig. 5-41 Step 15: Deposit and Define the Metal I Layer

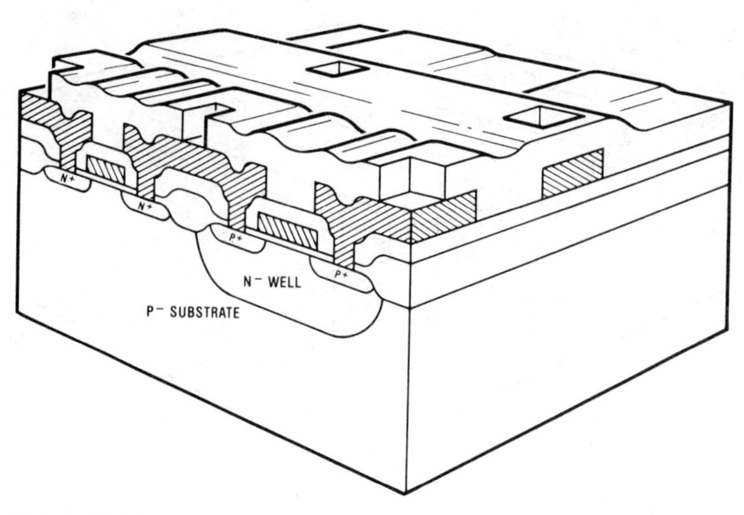

Fig. 5-42 Step 16: Deposit Insulator and Open Vias

vias, are then etched in this oxide layer to expose the underlying first metal layer at points which will allow electrical connection to the next uppermost (Metal II) interconnect layer.

In Step 17, Figure 5-43, aluminum is again deposited over the entire

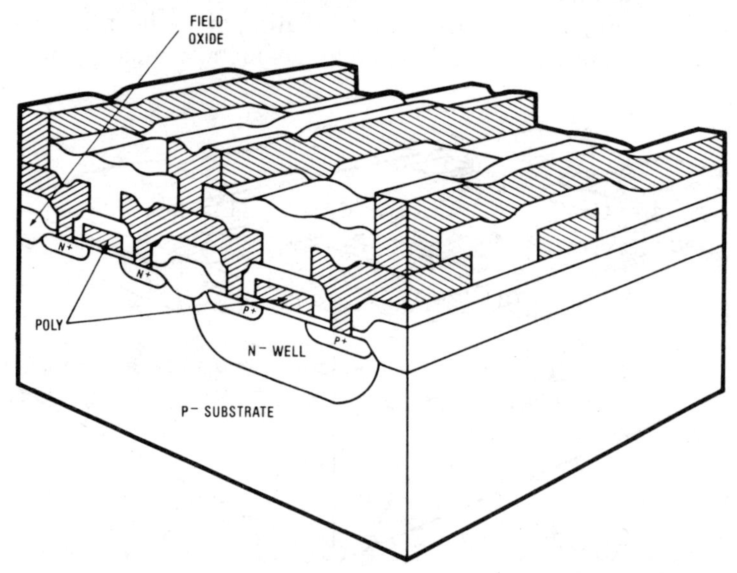

Fig. 5-43 Step 17: Deposit and Define Metal II Layer

wafer. Metal mask II is used to define this layer into the desired interconnect pattern. The bonding pads, for wire bonding to the package, are formed on this metal layer and this is the first time the completed circuits can be electrically tested.

In Step 18, Figure 5-44, another layer of thick oxide is deposited over the surface followed by an etch to expose the bonding pads. A final dielectric coating is then applied for further protection and is also etched to expose the bonding pads. The wafer is now ready for probe, electrical testing, backside preparation, thinning and then plating for die attach; sawing, and then packaging.

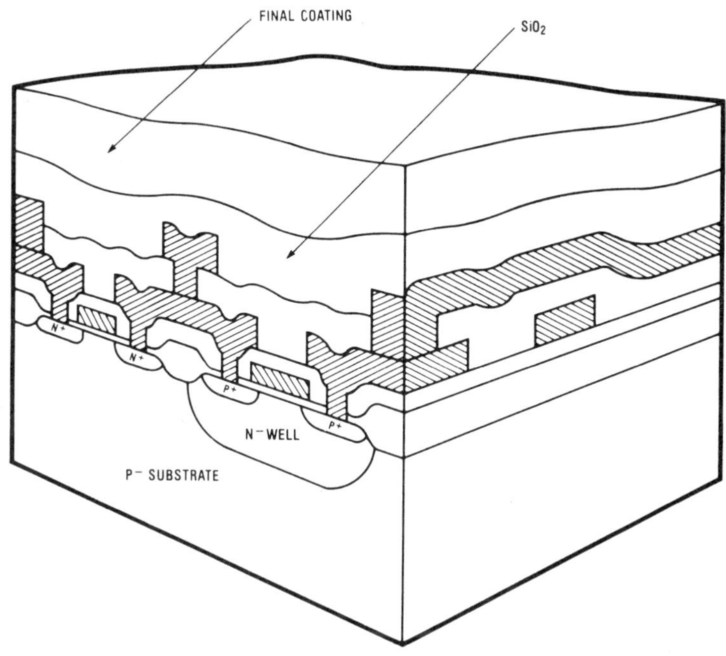

- DEPOSIT THICK OXIDE LAYER
- EXPOSE BONDING PADS
- APPLY FINAL WAFER COATING
- EXPOSE BONDING PADS

Fig. 5-44 Step 18: Deposit and Define Final Wafer Surface Coating

The benefits which result from this process include the following:

1. Much higher switching speed.
   A. The short channel lengths of the MOS transistors increase transconductance and therefore also increase the amount of current that is available to change node voltages.
   B. Smaller geometries reduce the parasitic stray capacitances.
   C. Self-aligned silicon gates reduce overlap capacitances.

   D. Oxide isolation eliminates the parasitic junction sidewall capacitance.
   E. N-well improves performance of the N-channel transistor and increases speed over a P-well process.
   F. Metal interconnects reduce resistance and the top metal layer reduces stray capacitance.
2. Low current drain.
   A. CMOS logic consumes power only when switched, so low-speed sections of a system reduce overall total power drain.
   B. Reduction of dc current flow greatly reduces the aluminum metal migration problem.
3. High reliability.
   A. Low junction temperatures result from low power drain.
   B. Reduced dc current flow improves reliability of metal interconnects.
   C. Special final surface passivation protects die.
4. Complementary transistors are available for combined analog and digital circuitry.
5. Smaller chip size.
   A. Results from small feature sizes.
   B. Old area-consuming guard-rings are eliminated with improved-performance oxide isolation.
   C. Multiple interconnect layers reduce chip area needed for hook up.
6. Wider range of CMOS logic voltage levels improves system noise immunity.

CMOS wafer processing is very dynamic. New additions and improvements are constantly being made. The above example is useful to indicate the complexity of the production flow.

CHAPTER VI

Semicustom and Custom Circuits

There are a number of techniques presently available to realize system logic circuitry. One very significant way to differentiate among these *Application-Specific* Integrated Circuits (ASICs) depends upon the number of IC fabrication masks that are unique to the circuit. If only the last few interconnect layers are "personalized," the approach is a *semicustom* circuit. If all the masking layers are personalized, it is a *custom circuit*. ASICs are expected to rapidly account for one-half of all the ICs sold. In this chapter we will provide some insight on this fast-moving area.

6.1 CMOS GATE ARRAYS, SEMICUSTOM ICs

Since the mid-60s, customers have had a growing interest in special, *customized* IC products that could be produced quickly and at low cost. Gate arrays are provided as one way to fill this need. These *semi-custom* IC products can be thought of as the next step in the evolution of complex chips for logic.

Logic designers have used Read-Only Memories, ROMs, to function as AND/OR logic devices. The organization of a ROM makes use of a large input AND array that provides for all of the possible logic combinations of the digital input signals. An output OR array is driven by this input array and is custom programmed to respond to a particular input code by providing a specific output code.

The basic idea of an AND/OR logic array is shown in the 2-bit (2-input) ROM of Figure 6-1. For clarity, this example is using diode logic. A ROM hard wires the Input AND Array to provide an output for every possible combination of the inputs. Programming a ROM involves customizing the Output OR Array to generate the desired specific output codes in response to each of the input codes.

Programmed Logic Arrays (PLAs) and Programmed Array Logic (PALs) are also derived from this same AND/OR array concept. The resulting name for the array depends on whether the Input Array and/or the Output Array is programmable or fixed-wired. PLAs are more flexible

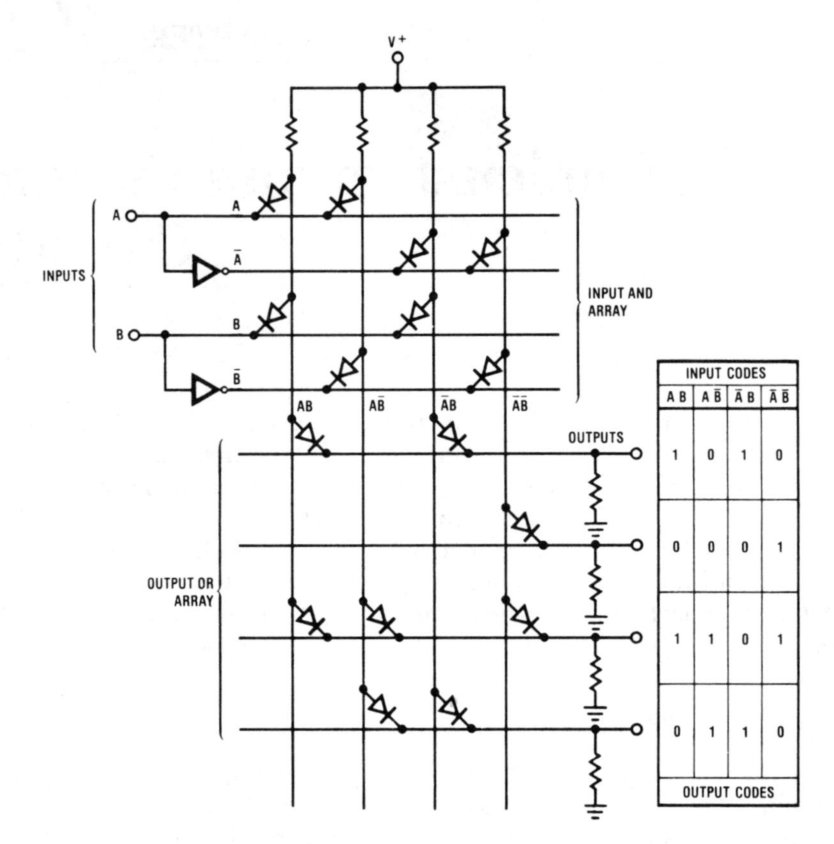

Fig. 6-1 Basic AND/OR Logic Array

than a ROM and they also allow more efficient silicon utilization because programming of both the Input AND Array and the Output OR Array is possible. The PAL circuits are the opposite of a ROM; that is, the Input AND Array is programmable and the Output OR Array is fixed.

The Philosophy and Uses for Gate Arrays

Today, there is even more flexibility for the system designer with gate array products. Compared to a custom circuit, gate arrays reduce risk, development cost, and turnaround time.

Gate arrays use a fixed assemblage of transistors, or gates, which can be configured into a logic block by the customization of one or more final mask interconnection layers in the wafer fabrication sequence. The total number of available gates is specified in the product name as, for example, a 2.4-K gate array. It is usually expected that at least 80% of these available gates can be "hooked up" or accessed in a given application.

The primary advantage of a gate array is that wafers can be almost completely processed, with the final *personalization* being relatively quickly done in the last few masking steps. An additional advantage of gate arrays is the possibility of qualifying a particular silicon chip. Logic variations can then be made on additional chips to suit many specific system requirements and these will also be approved because they use the same basic circuitry and identical wafer processing.

Gate arrays take the industry one step closer to the complete system on a single silicon chip. The objective is to combine all the logic of a large number of SSI and MSI packages which would normally occupy a large part of a PC board, and squeeze it onto a single gate array chip. It now becomes harder for your competitors to determine what is going on within your system.

System designers will exchange SSI and MSI logic IC catalogs for *macrocell libraries*, and PC board layouts will be exchanged for the placement and routing of interconnects on a silicon wafer. Circuit design is finally returning to the system houses.

By extensive use of software, system designers convert their logic diagrams into the detailed information that is needed to make IC interconnection masks. The semiconductor manufacturer will then make these masks and then pattern some existing, nearly finished wafers with these custom interconnection patterns.

Software-generated test tapes, assisted by the designer's inputs and timing requirements, will verify product functionality, and finished ICs will then be shipped to the customer—a cycle that is not much different from photographic film processing. A roll of film is sent in, and prints are returned. The same degree of responsibility sharing also exists. Blurred and poorly composed pictures are not the fault of the film processing lab; these are the responsibility of the *photographer*.

There are many benefits derived from the use of gate arrays. One of the more important is the reduction in the time that is required to design a new system. Performance advantages also result because only small-valued, on-chip parasitic capacitances must be driven as a result of the higher level of integration. This yields higher speed for the same power dissipation, or for the same performance as an SSI and MSI logic equivalent-function, a reduction in power consumption is possible. The smaller total number of external connections between components also increases system reliability. The lower power drain of CMOS allows more dense gate arrays as compared with the ECL bipolar gate array products.

Processes and Performance

In general, longer interconnect runs exist on gate array products. This has created interest in the benefits of multiple-layer metal processes, especially for the large-chip CMOS gate arrays. Early, double-metal processes with 3-μm feature sizes have provided single-gate propagation delays of 2 ns. A 2.4K CMOS gate array using this process dissipates less than 1W while operating and less than 10 mW during standby.

A major feature of the 3-μm CMOS gate array products was that the performance of the input and the output circuits matched both the speed, the ac specs, and the interface specs of the 74LS bipolar logic family. Also, the outputs met an 8-ns propagation delay spec when driving a 50-pF external capacitive load. The typical device performance parameters for this process are listed in Figure 6-2.

PARAMETER	N-CHANNEL	P-CHANNEL
V_{TH} (THRESHOLD VOLTAGE)	0.7 VOLT	−0.7 VOLT
BODY CONSTANT (M-FACTOR)	0.15 $V^{1/2}$	−0.43 $V^{1/2}$
β' (MOS g_m FACTOR)	32 $\mu A/V^2$	10 $\mu A/V^2$
BV_{DSS} (BREAKDOWN VOLTAGE)	9 VOLTS	−15 VOLTS
L_{EFF} (EFFECTIVE CHANNEL LENGTH)	2.37 μm	2.39 μm
V_{TF} (METAL 1-TO-FIELD THRESHOLD VOLTAGE)	26 VOLTS	−22 VOLTS

Fig. 6-2 Typical Device Performance for a 3μm, Double-Metal CMOS Process

Notice that the breakdown (the *reach through voltage*) of the N-channel transistor is 9 volts. This restricts the maximum operating power supply voltage spec to 6 V_{DC}.

The move to 2-μm channel lengths allowed CMOS gate arrays to provide a single-gate propagation delay of just under 1 ns. This is close to the 750-ps propagation delay that was achieved by the early ECL bipolar gate arrays. The power drain penalty of bipolar is demonstrated by the fact that the 1K ECL gate arrays drew approximately 10W of power.

Reducing the feature size to 1.25 μm provides the characteristics listed in Figure 6-3. The term *pitch*, used in this figure, is the total of the width of the metal, or diffused, line and the minimum width of the space that

L = 1.25 μm
W = 1.5 μm
CONTACT SIZE = 1.25 × 1.25 μm
M1 PITCH = 2.5 μm
M2 PITCH = 2.5 μm
GATE OXIDE THICKNESS = 200Å
FIELD OXIDE THICKNESS = 4000Å
N^+/P^+ PITCH = 3.5 μm
V_{CC} = 3V
LSI FIGURE OF MERIT = $5 \times 10^{11} \frac{Hz \cdot GATE}{cm^2}$

Fig. 6-3 Features of a 1.25-μm CMOS Process

must be used to the next allowed metal, or diffused, line. This particular 1.25-μm feature size process is seen to restrict the power supply voltage to 3 V_{DC}—a departure from the present 5 V_{DC} standard. This is the price that must eventually be paid for operation with short channel-length transistors.

There has always been a speed contest between bipolar and CMOS logic circuits. This performance evolution is shown in Figure 6-4. Bipolar

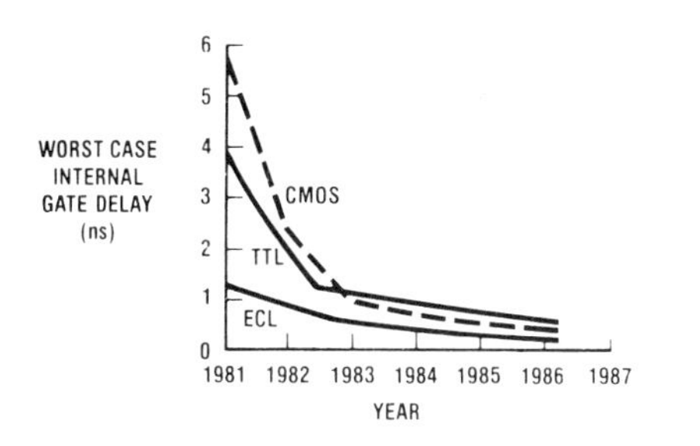

Fig. 6-4 CMOS Performance Evolution

logic is expected to win this race. Technologists have predicted an eventual average gate delay of 20 to 30 ps for bipolar logic circuits and 50 to 100 ps for CMOS logic circuits.

The floor plan of an early 2.4K CMOS gate array is shown in Figure 6-5. There are input buffers along each vertical edge of the chip, and I/O

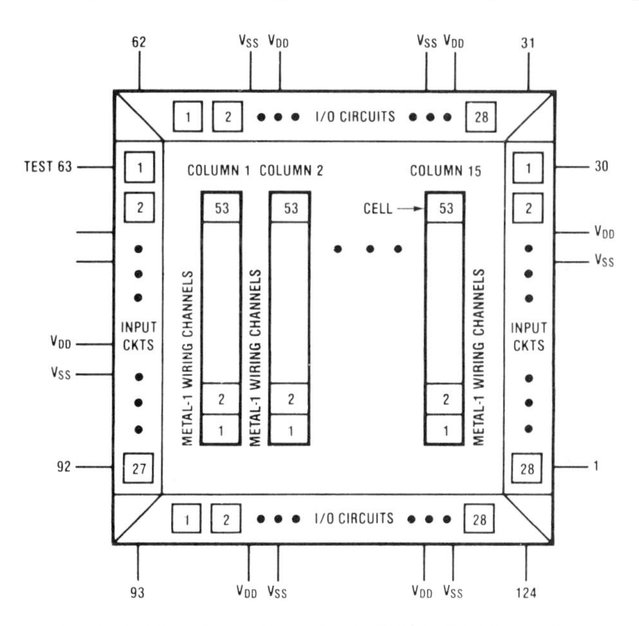

Fig. 6-5 The Floorplan of a 2.4K CMOS Gate Array

buffers along the top and bottom edges. These I/O buffers can be configured as either additional inputs or as outputs for the array.

The internal groupings of the 16-transistor cells, 8 N-channel and 8 P-channel transistors, are arranged in a 15 × 53 inner matrix. The spacings between these cells allow for the metal routing lines. Note how the photomicrograph of the chip, Figure 6-6, follows the floor plan. The I/O buffers

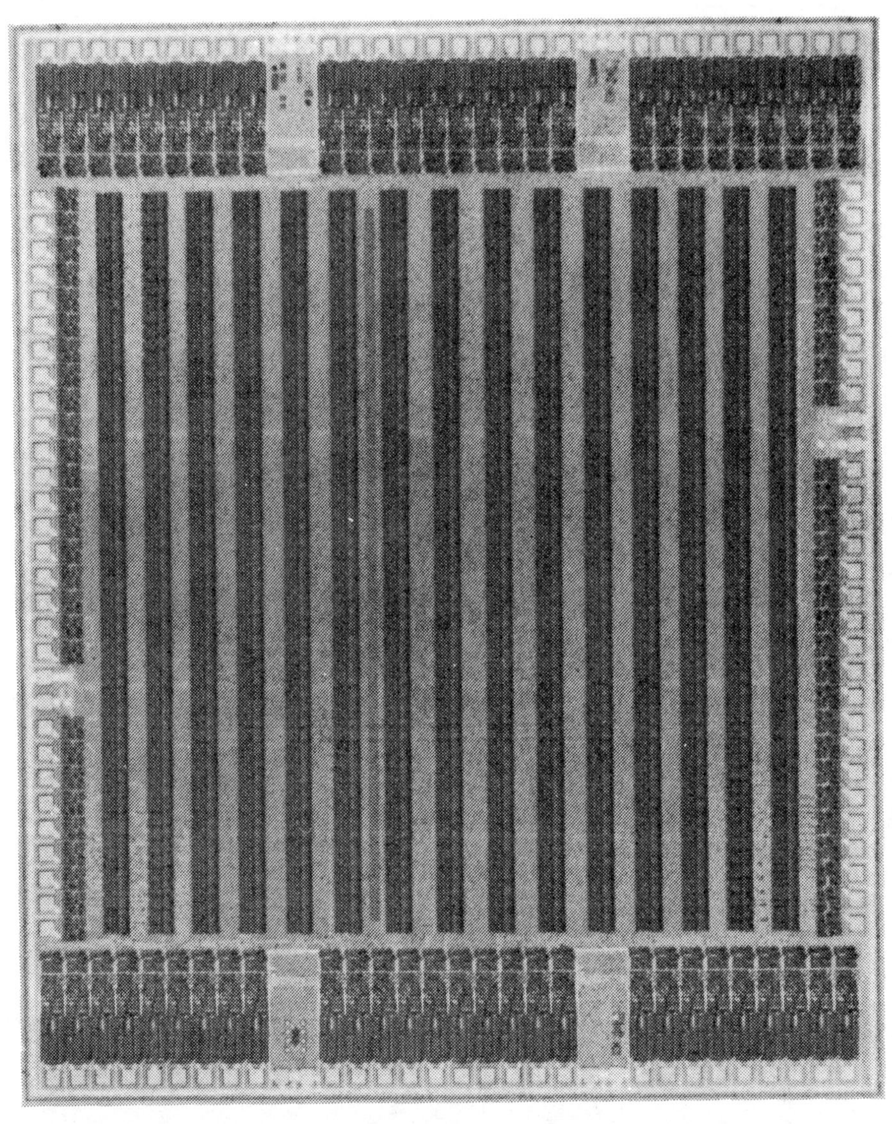

Fig. 6-6 A Photomicrograph of a 2.4K CMOS Gate Array

along the top and bottom edges are seen to be considerably larger than the input-only buffers along the vertical edges.

When CMOS gate arrays reached the complexity of 6K routable gates, additional circuitry was added on the chip to ease the testing problem. This extra circuitry was called the On-Chip Test and Maintenance System (OCMS). The die organization of an early 6K CMOS gate array is shown in Figure 6-7. The OCMS is shown across the bottom and involves a pseudorandom number generator that provides the input series of codes, the *input vectors*, for the test. This generator is initialized at the start of the test and will then always repeat the same code sequence.

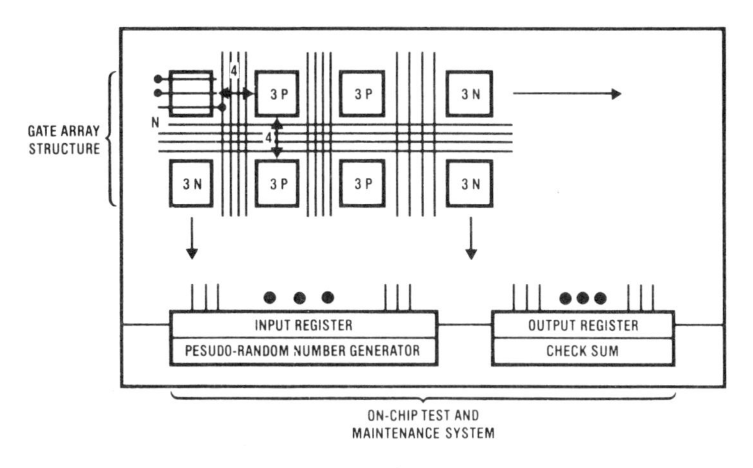

Fig. 6-7 Die Organization of a 6K CMOS Gate Array

A register at the output is also reset at the start of a test. This register sums the output codes that result from the random stimulation of the inputs. Performance verification is achieved by comparison of the final result in the output summing register with a previously-stored correct check sum. This chip had 172 pins: 66 inputs; 88 I/Os, can be used as either inputs or outputs; 12 for power and ground and 6 for testing.

Notice, in Figure 6-8, that each of the three transistor cells are all of the same type, N-channel or P-channel. The horizontal and vertical spacings between the transistor groups are for the routing lines. The dotted lines shown on this figure are the poly gates for the MOS transistors. The source and drain regions of these transistors can be interchanged, paralleled, or the transistors can be placed in series to obtain a desired logic cell.

CMOS gate arrays are expected to continue to increase in density with an eventual limit of 500K to 1 million gates.

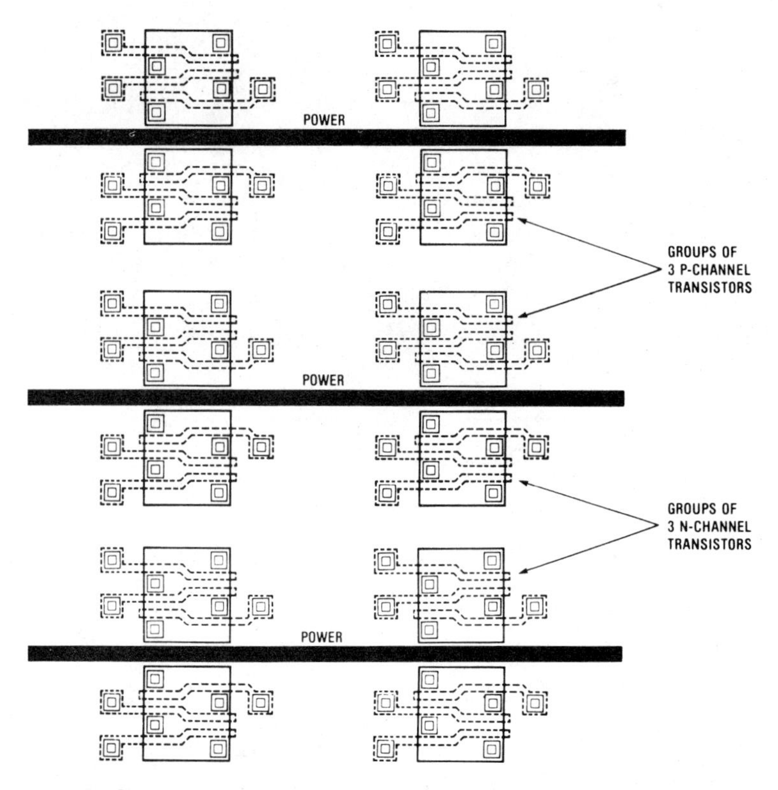

Fig. 6-8 The Details of the Transistor Cells on a 6K CMOS Gate Array

The performance of all of the CMOS gate arrays generally follows the same rules of thumb that were mentioned for the 74HC CMOS logic family: propagation delay increases with increasing ambient temperature at the rate of approximately 0.38%/°C and the propagation delay also increases as the power supply voltage is reduced at the rate of approximately 2% per 100 mV of power supply voltage decrease.

The unusually large number of inputs and outputs of gate arrays has forced the use of packages with a large number of pins. (We will consider some of the developments in IC packaging concepts in the next chapter.)

Computer-Based Design Aids

The major activities involved during the application of a gate array product are shown in Figure 6-9. All of the steps necessary to do a PC board layout are now done by computer programs. For example, if you were to design a PC board that realized a given logic diagram, you would first search the SSI and MSI logic databooks to find a minimum number of IC packages that would perform the given overall logic function. Next, you would use

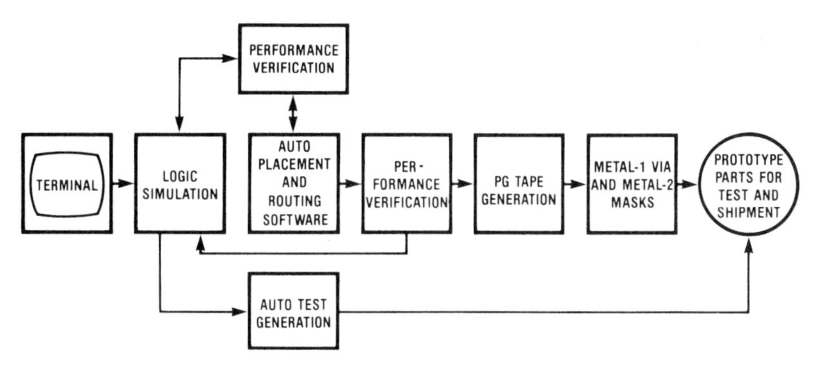

Fig. 6-9 A Gate Array Design Automation System

trial and error iteration to locate, or *place*, all of these packages on the PC board in a manner that not only reduced the lengths of the IC interconnect wiring, but also would allow placing the IC packages in an orderly array on the PC board. If the number of interconnect wires, *routing* wires, for this *placement* became excessive in a given localized area, you would then try a different *placement* of the IC packages to solve this *routing congestion problem*.

Computer programs must now automatically convert from a logic diagram to the selection and layout of actual hardware. Whenever possible, the pin-outs of a gate array should not be preassigned. This significantly simplifies the layout job.

Most of the current design procedures for gate arrays start with a specification of the desired logic to be implemented. This logic diagram can be entered into a computer in many ways. Unfortunately, the simplest one to program makes use of a line-by-line specification of the logic gates and the required interconnections (a netlist) and this is the most laborious for the human operator. Netlists require that all of the inputs, outputs, and internal logic gates be given a descriptive name. Then each interconnection of the logic diagram must be specified in a way that indicates which of the internal nodes are to be interconnected.

An easier logic entry concept, from the human operator's viewpoint, is *schematic capture*. Sophisticated computer programs allow essentially a direct computer entry of the logic diagram. This is a much easier and a significantly less error-prone process.

Following the entry of the logic diagram into the computer, the next step in a computer-based gate array design is a logic simulation to verify the data entry and to insure proper logic functioning of the array. During this phase, test programs must also be generated and *fault graded* to insure adequate detection of bad chips.

Fault grading is accomplished by computer logic simulation that sequentially sticks each internal circuit node at both a logical "1" and then a logical "0." The computer then determines if the intended testing sequence will detect these internal simulated faults. The percentage of these faults that are detected by the testing program provides the numerical *fault grade*. A fault grade of at least 85% is generally required, and most array applications provide much higher fault grade results. To improve a fault grade, additional tests have to be added or the on-chip logic must be altered to make it more testable. This testing problem is a major part of a gate array application and often requires the use of additional circuitry to achieve the testing goals.

Once a proper logic diagram has been determined, including the testing requirements, the next step is a trial placement or location of a selected set of macrocells, SSI/MSI logic blocks, that will provide the logic needed. Most placement programs require human assistance and are therefore *semi-automatic* programs. A first trial placement is based on the required interconnection between macrocells.

After a placement has been determined, a *stub router* subroutine is then called that indicates the connections, or *stubs*, that are needed for each cell. A *maze-running router* subroutine then attempts an interconnect, the routing step. If a routing congestion problem results, a *rip-up and retry* subroutine is called and a different placement of the cells is made in an attempt to relieve the routing congestion. The goal of this software is to guarantee routing if only 80% of the existing gates are used in a given gate array design.

Routing algorithms trace back to 1959 when E.F. Moore of Bell Labs developed the *Moore Algorithm* to route telephone calls through a switching network. Moore's work was extended to a two-dimensional grid by C.Y. Lee, also of Bell Labs, in 1963. This *Lee Router Algorithm* is the basis of most of today's *maze routers* that interconnect the macrocells. Lee's work was augmented by Hightower, also of Bell Labs, in 1969 to provide the *line-search algorithm*. This takes up less computer space and executes from 10 to 15 times faster than Lee's algorithm.

In 1971, a new type of router was developed, the *channel router*, by A. Hashimoto and J. Stevens of the University of Illinois. This was a very efficient method when the points to be interconnected were arranged in parallel rows separated by an interconnect channel. This is why the channel router has been applied to gate arrays. When the allowed channel width is not wide enough to accept all of the required interconnect lines, the *channel density* is said to exceed the *channel width* and either a new routing, a new placement, or a wider channel must be used to solve this routing congestion problem. A good router still cannot make up for a poor placement of the macrocells.

The use of large logic blocks, the macrocells, aids the placement problem. Placement optimization results from the use of *pairwise interchanging*, the interchange of two cells, to determine if a reduction in the total interconnect metal length is achieved. This quickly becomes a large computer problem because each trial interchange must also be trial routed.

There was no lack of individual computer-based tools to solve the automated array design problem. What was lacking was an easy way to tie all of these independent software tools together, because they were not initially developed to work together. The development of gate array design automation programs has taken *hundreds* of man years; and as we move on to new, more complex approaches for semicustom circuits, this gate array effort will help, but there are still large problems remaining in providing a totally automatic design system.

Notice in Figure 6-9 that once the placement and routing is completed, a check of the logic that exists on these masks is made to insure that the original logic design has been properly implemented. This is known as performance verification.

A transient analysis of the performance of the complete array can be computed when all of the additional capacitive loading of the interconnect lines is accounted for. Transient analysis programs show the actual timing, to within approximately 10%, of the finished product.

Once the final verifications are complete, pattern generator (PG) tapes are made that allow the fabrication of the masks: the first layer metal, the second layer metal, and the via mask, for connection between the metal layers. These masks are then used to pattern this custom logic onto existing silicon wafers. Finished parts are then tested and shipped. This oversimplified sequence of steps completes the gate array application design cycle.

High-performance computers are required to support this design cycle and advanced graphics systems are also needed. Even the comparative simplicity of this example of the first of the semicustom alternatives, the gate array, has been a large computer-software design problem.

Future Possibilities of Gate Arrays

Will gate arrays take over and be used for everything? Not yet, but in the meantime, tough competition for gate arrays will result in cost advantages for the customer who purchases handcrafted, high-volume, product-specific IC circuits.

The challenge facing the gate array product groups is to find ways to increase the circuit count but to limit the chip size. To achieve this goal, researchers have reported the use of four levels of metal to significantly increase circuit density in bipolar gate arrays.

As we look into the future, we see an ever-increasing testing problem that may limit the density of CMOS gate array chips. This has demanded the inclusion of the testing factor in the chip design phase.

The demand for linear functions, to allow single-chip complete systems, has forced the development of combined linear/digital arrays. In addition, configurable RAM and ROM memory blocks have been incorporated to increase the usefulness of gate array products.

Software is the key to solving the complexity problems, reducing human errors, and reducing the turnaround time. Large expenditures in software have been made, and this can be expected to continue to make it easier to obtain products.

Gate arrays, like the microprocessor in the past, are more than just another IC product. Heavy investments are required to provide the service the customer needs so he can build the new complex-systems of the future. As each year goes by, it costs more to stay in the IC manufacturing game, and there's more to it than just processing silicon.

Gate arrays are dynamic. They will follow the customer's lead. The eventual complexity of CMOS gate arrays is expected to reach a limit at about 0.5 to 1×10^6 gates. The ever-increasing amount of circuitry on a single chip continues to alter the thinking and the planning for the architectures of future systems. (There will be more on this in Chapter 8.)

6.5 CUSTOM CIRCUIT ALTERNATIVES

Designers of systems that will be in high-volume production can usually justify custom IC chips. Custom circuits customize all of the masking layers and therefore provide more efficient utilization of the silicon chip in exchange for higher tooling costs and longer turnaround times.

Moving in a direction of increasing design time, there is a sequence of chip design approaches that starts with gate arrays, then proceeds through many custom alternatives such as: standard cells, originally called *standard-height cells*; *general cells*, variable height and variable width cells with the addition of ROM, RAM, and PLA; and finally *silicon compiled layouts*.

Through this sequence, the chip area for a given logic function will generally decrease, but the design cost will increase. The optimum choice for a given system depends on the present state of IC technology, tooling costs, total number of chips needed, time available for completion, and the possibility of short-term design changes. The selection of a design approach becomes very difficult because many of the important inputs may not be known at the time the chip design must be undertaken.

Some observers have noted that existing digital systems usually have a certain percentage of unused circuitry within the standard IC packages. Therefore, it has rarely been the case that a system has had optimum utilization of the available silicon area. These spare circuits have often come in handy to support the last-minute system changes that always seem to come up just prior to shipment. This ability to respond rapidly to a change is an important consideration in any proposed new system.

Standard-Cell Arrays

Standard-cell arrays are defined to be fixed-height library cells that are used in combination with variable-height routing channels. In this case, the silicon is not preprocessed and all layers are customized. New circuit designs are also not attempted, so this is very similar to the standard PC board layout of standard SSI and MSI logic circuits.

This approach allows more complex cells to be used and often results in a factor of two speed improvement over gate arrays. There is a more efficient utilization of silicon because all layers are customized, and the chip area can often be reduced by a factor of one-half that of a gate array. Routing congestion is eliminated as a result of the use of variable-width routing channels. And, finally, the use of structured design with this approach allows fast design time.

A disadvantage of many approaches to cell arrays is that they do not support ROM, RAM, PLA, and other logic blocks. This can become a major limitation in memory-based architectures. Also, the fixed-height cells usually prevent using larger, more efficient blocks of logic.

A limited improvement over standard cell arrays is provided when a few, large, special-purpose blocks, such as memory, can be added within the cell array, as shown in Figure 6-10. This now supports memory-based architectures, but it is not general enough to support more than one or possibly two of these special-purpose blocks.

Advantages can be gained by making use of *general cells*, which make use of both variable height and variable width cells. Also ROM, RAM, and PLA can be added to these cells. This is the next level of semicustom IC chip complexity.

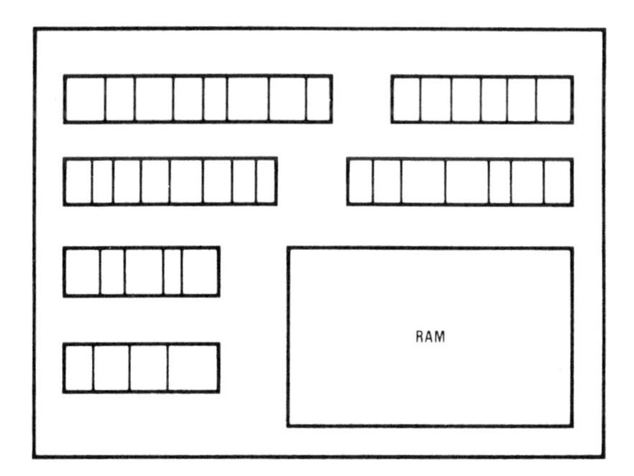

Fig. 6-10 A General Floorplan of Standard Cell with Memory

Functional Block Arrays

A more general approach has been called *Functional Block Array* (or *Macro-Cell Array*). This allows large rectangular blocks of circuitry of any size to be used: RAM, ROM, PLA, ALU, A/D, DAC, filter, or amplifiers, as shown in Figure 6-11. All of the standard cells can be treated as predesigned blocks. This is very much like a *block* diagram of your system. Special software tools support building up larger systems out of these simpler blocks.

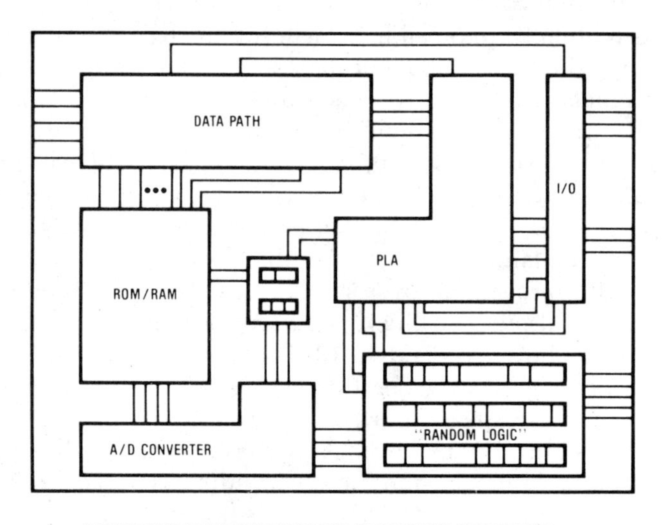

Fig. 6-11 A General Floorplan of a Functional Block Array

With this approach, very silicon-efficient circuit blocks can be made available in a circuit library. The routing requirements for this functional block approach are also well within the state of the art.

Silicon Compilers

The ultimate way to approach a VLSI chip design is to make use of a special, high-level computer language in which all of the attributes of a new system can be specified. A *silicon compiler* then is used to translate this high-level system description into the architecture, circuits, and layout details that are necessary to produce an IC that meets the specified goals.

A problem of inefficient use of the chip area exists with layouts made with a silicon compiler, and this therefore tends to restrict silicon compilation to use by companies that have a large interest in low-volume custom circuits where the reduction in the design time is a larger benefit than the increased cost that results from a larger chip size. Unfortunately, the proprietary nature of these design tools means that the specifics are not generally available. High-sales-volume ICs are expected to continue to make use of more expensive, more handcrafted chip layouts.

CHAPTER VII

IC Packaging Developments

The gate array products have forced the IC industry to find ways to economically package high-leadcount, 68 pins and greater, products. The largest pin counts will be found on the custom and semicustom ICs and projections of 500- to 1000-pin packages with power dissipation on the order of 100 watts are being made. New packaging solutions, each with a different acronym, are rapidly appearing, but no standards are imminent. This requires both the IC suppliers and the IC users to remain flexible during this rapidly changing package-development phase.

A combination of both *insertion* and *surface mount* packages are appearing and both are available in ceramic for military applications and plastic for industrial and consumer applications. (It is interesting to note that approximately 95% of the ICs shipped are in plastic packages.)

The increasing use of high-density component packages, with greatly reduced terminal spacings, is causing a shift away from the traditional insertion-mounted packages to new surface-mounted packages. The reduced lead spacings favor the elimination of the requirement for holes in the PC boards. Surface mount packages are directly soldered to the PC copper traces on the boards.

The soldering operation is accomplished by reflowing a solder paste. The required 200°C heating can be supplied in many ways: via a vapor-phase reflow system; convection or infrared furnaces; or by pulse reflowing, where a pulse-heated tool makes thermal contact at the solder joints. The more traditional wave soldering operation is typically limited to the attachment of passive components and small-outline packages.

The typical electronics engineer is unaware of the difficulty of the packaging engineering problems. Many conflicting requirements must be simultaneously addressed: thermal performance (including the thermal coefficient of expansion of the materials used in the package and that of the PC mounting board material); electrical performance, lead self-inductance and also the stray inductive and capacitive lead coupling are problems in large-pin-count packages; cost per lead; test-ability, via automatic systems; reliability, must provide die protection in many hostile environments; hermeticity, leak-proof requirements; PC board compatibility; and density. All of these factors must be simultaneously considered in package design.

Continuous improvements in both *quality,* those factors that guarantee that all of the ICs that are received by a customer meet the required specs, and *reliability,* those factors that extend the life of the IC once it is placed in service, are being made. The present goals have been met by simultaneously incorporating many changes in all aspects of the manufacture of the modern IC products. For example, continuous research is in progress to provide a *bullet-proof* die by adding additional protection to the die surface during wafer fabrication to reduce the susceptibility of performance degradation resulting from packaging contamination. There is also continuous development to improve plastic molding compounds and to improve both lead-frame materials and lead-frame design. Quality and reliability considerations are also factored into the IC circuit design phase. Multi-pass final testing, and the addition of increased automation; computer control, and reporting on the wafer fabrication production lines are all contributing to improvements in both quality and reliability.

Packaging engineering groups today provide personnel with expertise in many disciplines: chemistry, both polymer and analytical; metallurgy; mechanical engineering; physics; and electrical engineering, including software. These interdisciplinary engineering groups are solving the customer's packaging needs for increased system density, performance, and reliability while reducing the system costs.

The new high-density packaging techniques have reduced the unit cost per system by up to 50%. First generation density improvements have achieved a 3:1 density increase and it now appears that a 6:1 increase in density is possible. The resulting smaller-sized systems are more reliable because there are fewer PC boards and also fewer cardedge and other connectors.

Electronic systems are generally either *cost* driven or *performance* driven, but sometimes both of these considerations apply. The current move to *high density packaging,* which improves the density that can be obtained over the standard dual-in-line IC packages, is solving both of these problems. For example, it has been shown that the relative system cost decreases as the square-root of the physical sizes, the volumes, of two alternate system packaging approaches. Reducing the volume of a system by a factor of four reduces the unit system cost by approximately a factor of two. Further, the physical size of a system affects the interconnection propagation delay. This time lost in communicating between the various circuits of a system can account for one-third to one-half of a critical-path delay. For well-partitioned systems, this critical-path delay is roughly proportional to the square-root of the area of the PC board space used. If high density packaging (achieved by the use of low power-drain

CMOS circuits) can reduce the required PC board area by a factor of four, then the resulting factor of two reduction in the critical path delay can often provide a 20% to 30% increase in the overall system speed and can therefore command an increase in the selling price of the system.

7.1 HIGH-DENSITY REPLACEMENTS FOR THE STANDARD PLASTIC DIP

There are three high-density replacements for the standard 0.100 inch lead spacing, plastic, dual-in-line package (DIP): The *Small Outline* (SO) packages shown in Figure 7-1; the Single Inline Package (SIP), shown in Figure 7-2, and the Plastic Chip Carrier (PCC), shown in Figure 7-3.

The SO packages are currently in high volume production, use a 50 mil (0.050 inch) lead spacing, and are available in both a narrow (N),

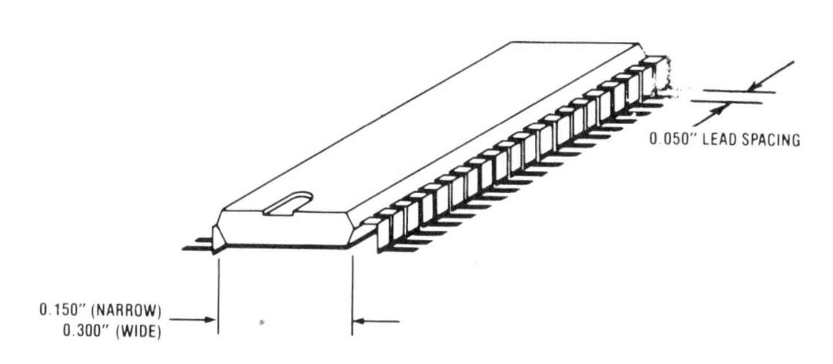

Fig. 7-1 The Small Outline (SO) Package

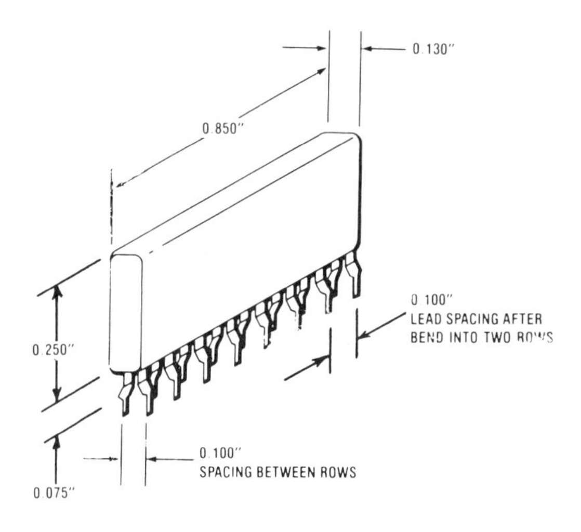

Fig. 7-2 The Single Inline Package (SIP)

Fig. 7-3 Plastic Chip Carriers (PCC)

0.150-inch wide, and a wide (W), 0.300-inch wide, version. These packages cover the range of from 3 (the Small Outline Transistor, SOT) to 28 leads.

The SIP package, which has been introduced by some Japanese consumer IC houses, is a molded package and mounts on one edge into a PC board. It has approximately a 2:1 overall density increase over DIPs and works best with forced air cooling. The popularity of this package is questionable, although it is also being considered for an IC memory product. This uncertainty of widespread acceptance is delaying the development effort.

The PCC packages also use a 0.050-inch lead spacing, but provide a 3:1 density improvement over the DIP. They are better for multi-leads and are now the industry standard for 68 leads. This package is expected to soon account for 35% of the total market. A wide range of leadcounts is projected: 16, 20, 28, 44, 52, 68, 84, and 124. All of these DIP replacements use existing packaging technology. The same PC board, the same solder-reflow technology (screen on solder paste, attach units, and use a vapor-phase solder reflow), and the same equipment (the same automatic screen printer, the same pick and place—tape or stick-machinery—and the same vapor-phase solder reflow system) can be used. Also, all of these packages are cost effective.

There is strong interest within the industry to reduce the lead spacing to 0.025-inch in a standard, high-density, low cost package. In addition, there is a move to the use of preplated (solder plated) lead frames that can retain good solderability through the packaging flow. This will increase the reliability because it will remove the solder immersion step that is normally used at the end of the line and will also provide a controlled, uniform thickness of solder on the leads.

For military and telecom applications, ceramic high-density packages will replace the PCC and SO packages. One of these, the *Leadless Chip Carrier* (LLCC), shown in Figure 7-4, will handle 16, 20, 28, 32, 40, 44, 52, 68, 84, and also a special 124 lead option (for gate array products). This package is somewhat hard to handle and is therefore not widely used. A variation of this package includes the leads and is therefore called a *Leaded Chip Carrier* (LCC).

A very high density *Cerpac Chip Carrier*, shown in Figure 7-5, is also being developed. This package uses 0.025-inch lead spacing. These packages will follow the PCC and LCC footprints and will provide cerdip/cerpac quality and cost.

7.2 PIN GRID ARRAY PACKAGES

Another high density package is the *Pin Grid Array* (PGA), shown in Figure 7-6. These *bed-of-nails* packages use a ceramic substrate, have 0.100 inch pin spacings, have excellent thermal characteristics, and provide a very high density; but, unfortunately, they are rather expensive. Ceramic PGAs exist with 68 to 308 pins.

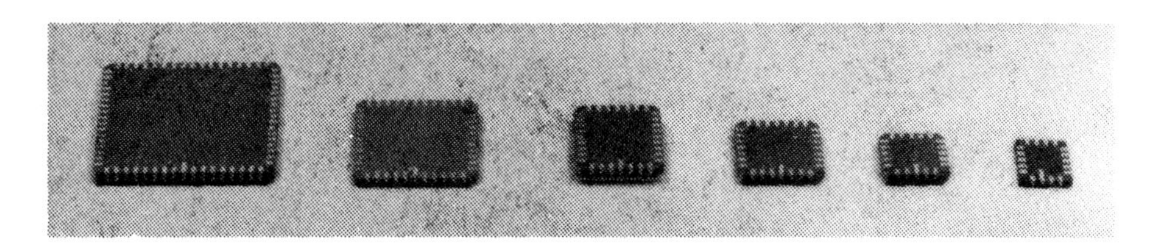

Fig. 7-4 Leadless Chip Carriers (LCC)

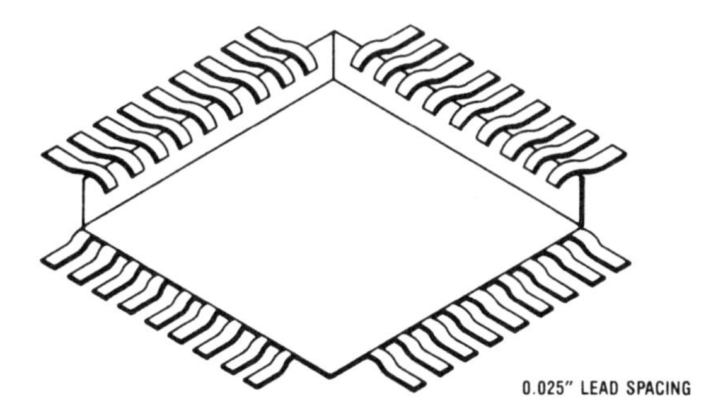

Fig. 7-5 The Cerpac Chip Carrier

Fig. 7-6 The Pin Grid Array (PGA)

7.3 TAPE AUTOMATIC BONDING AND TAPE-PAK

To improve reliability and to reduce the cost of producing multi-lead ICs, a single gang-bonding is desirable over the standard multi-leadwire bonding. This usually requires special bonding *bumps* to be placed either on the bonding pads of the IC chips or a *bumped tape* must be used. The bumped tape has the advantage that less special die preparation is required so costs are reduced.

Tape-automated bonding has also been applied to a high-density surface-mounted IC package that has been called *Tape-Pak*. This package, shown in Figure 7-7, utilizes one continuous tape lead from the die to the PC board and also includes a protective enclosure for the die. It is planned to be available for both military applications (H-pak) and indus-

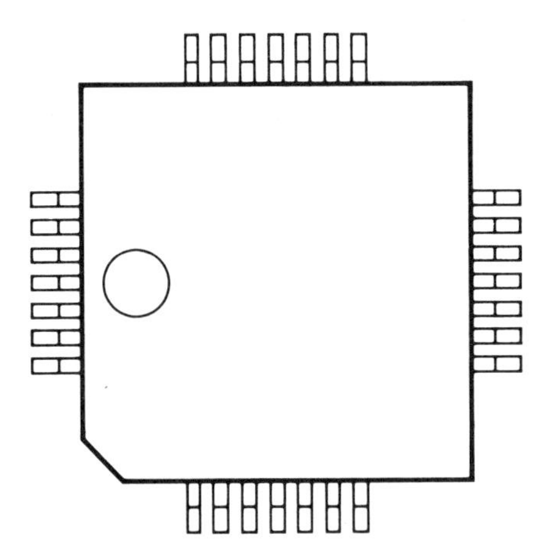

Fig. 7-7 The Tape-Pak Package

trial/consumer applications (P-pak). This package is designed for surface mounting, which implies that the package leads will extend radially from the package and overlay onto a matching copper pattern on the PC board. Three lead-configuration possibilities are shown in Figure 7-8.

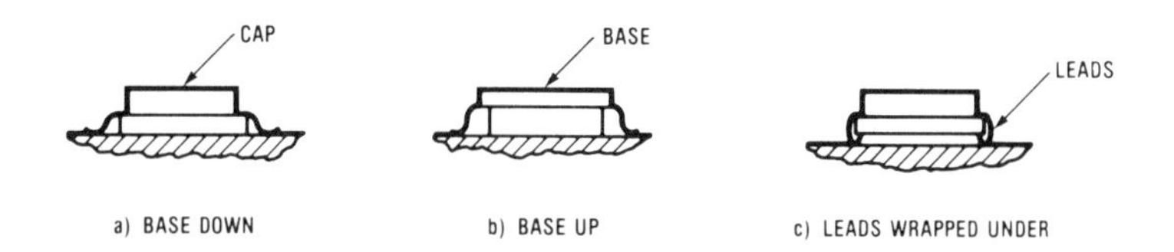

Fig. 7-8 Tape-Pak Lead Configuration Options

A comparison of this Tape-Pak with other packaging types is shown in Figure 7-9. Both the system cost factors and the system performance factors have been normalized to the small, 0.0166 inch lead spacing, Tape-Pak. Note that the projected system cost reduction factor is 2.5:1 and the system performance (speed) improvement factor is 34% when compared with the 0.050 inch lead spacing PCC package.

The packaging of semiconductor chips has always been important and may set the limits on the computers of the future unless these future computers become single-chip systems. The problems of cooling the chip, providing power, protecting the chip, and getting the signals from package to package—while providing new standards of miniaturization—become

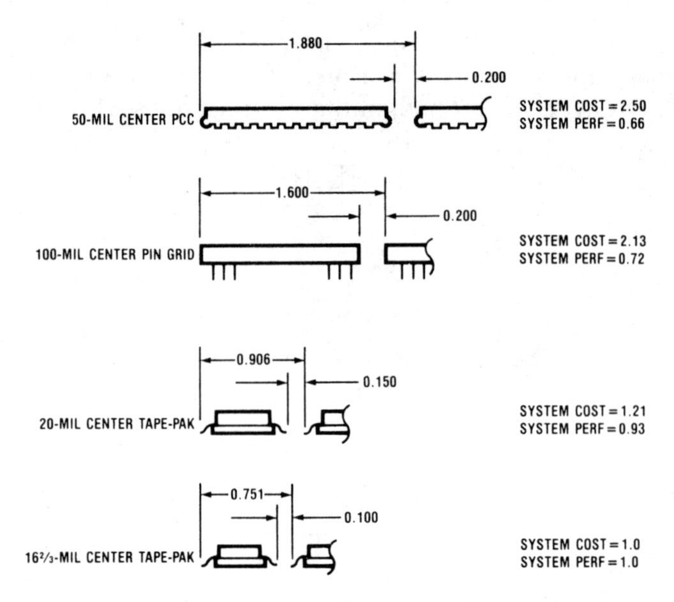

Fig. 7-9 Comparing Tape-Pak with other Packages

very challenging. There is much activity in this area and significant resources are being applied to provide solutions both for the high-performance, high-speed systems and also for the more cost-conscious applications where low-cost per interconnection is an important consideration.

CHAPTER VIII

A Look into the Future

There is an increasing need for the IC suppliers and the IC users to work together more closely on items such as quality, reliability, incoming inspection, and the design and production of Application-Specific ICs (ASICs). Different customers will need VLSI design help to be supplied in different ways. Some customers will require only software packages; others, terminal access; and still others, stand-alone design centers. Software support is a key ingredient of service.

The processing of silicon is rapidly racing to the ultimate limits predicted by the technologists. The major limits of VLSI are not expected to be based on the physical laws of nature. More practical considerations have been predicted to dominate: the reduced reliability that results from extensive device size-reduction, the limited economic resources that can be made available to provide the necessary computer-based tools, and the increased costs and the reduced throughput of the new VLSI fabrication equipment. Many observers are also concerned that a move away from the 5V standard power supply voltage will meet with much opposition. But this change is mandatory if the smallest channel-length transistors are ever to be used.

There is a trend toward using a complete wafer for a large memory or a microprocessor function. A major feature of these designs is the heavy use of redundancy to guarantee that the finished wafer will be functional. An important advantage of this approach is the high speed that results because of the reduced path length and the reduction in off-chip signalling that is used.

8.1 DEVELOPING THE SINGLE-CHIP VLSI SYSTEMS OF THE FUTURE

The VLSI age is upon us, and researchers are hard at work on Ultra-Large-Scale Integration, ULSI, (gatewidths less than 0.1 μm and 0.1 μm linewidths) and Super-Large-Scale Integration (SLSI), yet the question remains whether VLSI is for everyone. For example, will the smaller electronic system companies be able to afford the relatively high front-end tooling costs of VLSI, and will the lack of VLSI components in a product reduce its competitive edge and put some companies out of business? These questions will be answered in the coming decade. The low-cost systems that will result from the VLSI era will increase the overall demand for silicon products, which will be a major benefit for the semiconductor companies. This increased demand for complex ICs will more than counteract the decline in the total number of packages that are shipped.

ASIC chips are the only way to take advantage of the increasing IC circuit complexity. There are just not enough complex products, other than μPs and memory, that will simultaneously suit many different users. We must not simply find ways to implement existing TTL-based logic diagrams—we now have the added benefits that result from architectural freedom.

The major benefits of VLSI can be achieved when system partitioning that minimizes the pin access of each chip to the external world is selected. It is interesting, that as more logic is placed on a single IC chip, the pin count at first increases. As even higher levels of integration are reached, the pin count turns around and decreases.

The Motivation

The economic benefits of placing more of the electronics in one package were even realized in the vacuum tube days. This resulted in the appearance of dual-triodes and other multi-function tubes. The Transistor Age also responded to this economic pressure by phasing into the IC Age. Today, we are approaching the point where it may be feasible to place all of the active, passive, logic, memory, and linear functions that are required for a wide range of systems onto a single chip.

There are also many additional system advantages. The small physical size of the silicon chip reduces parasitic circuit stray-capacitances. This increases speed and/or reduces power consumption. The reduced *cost* of added on-chip circuitry allows new VLSI designs to include extra circuits to solve the increasingly complex testing problem and also adds the possibility to use extra circuitry to automatically reconfigure the chip in response to a chip fault condition. In addition, more complex system architectures are possible—to either enhance system performance or to solve other system problems by use of more complex IC electronics.

The New System Architectures

VLSI allows memory and logic to be placed on the same chip. This relatively simple statement has profound implications and is the reason for the current upheaval in the area of system architecture. The theoretical question that now arises is how to best take advantage of this new VLSI electronic hardware breakthrough.

New words are appearing, such as *highly concurrent* systems. This system idea is to allow a large number of calculations and/or processes to take place at the *same time*. This is in contrast with the *sequential systems* of the past where problems were solved by stepping *sequentially* through program steps that constituted an algorithmic solution to the system problem. When we consider the complexities of these inadequate, *old-fashioned* sequential systems, the problem becomes how to optimally design these new mind-stretching concurrent systems. This challenge has stimulated the academic circles and much current research is focused on this problem.

System designers must discipline themselves so that they can get away from the traditional thinking of problem solutions as part of a sequential machine. The irony of this is that it wasn't that many years ago that all logic designers were told that the *new* thing was to stop thinking about problem solutions as interconnections of 7400 TTL gates, and to learn to attack problems so that they could be solved on sequential machines! This has caused the conventional sequential machine to have a strong grip on current thinking. (You can see the problem of being a digital system designer. Every few years someone is telling you that what you are doing is no good—and that there is a better way—but this better way is quite different, unfamiliar, and takes unavailable time to get used to.)

The flexibility and the regularity of sequential architectures is still found to be effective in many special purpose IC chip layouts. This regular layout, the ability to easily change algorithms and even to relatively easily handle the chip testing problem, via external access to the internal data bus, makes this approach very useful for many smaller IC chips. However, where speed is important, other architectures usually have an advantage.

Everyone agrees that the use of random logic should be avoided. Chip layout errors are easily made with random logic designs and changes are also very difficult to implement. Structured logic design is therefore becoming important and this tends to produce a large degree of *regularity*, the use of common layout cells, in the chip design.

Also, dynamic logic designs can reduce power drain, even in CMOS systems, but unfortunately introduce critical timing problems so testing becomes a larger problem. This has favored the design of systems that use static logic.

When it is economic to increase circuit complexity to improve speed, *pipelined* architectures are used. This idea is used on any production line. The key is to design all of the timing of the supporting activities such that the flow of the main problem is never held up.

Multiprocessing architectures are also possible where each of the processors can communicate with the other processors, much like people do in a business organization. It has been speculated that pipelining plus multiprocessing at each stage of a pipeline will lead to the best possible system performance.

All of the problems that have existed with the large digital systems now can be found on a silicon chip. For example, the problem of *clock skew* occurs on current IC chips. Clock skew occurs when the clock does not *strike* simultaneously in large digital systems owing to the finite propagation delay of electrical pulses as they propagate down wires.

Various solutions to this problem have been proposed, such as a *self-timed* architecture where computational assignments are made to auxiliary processors and then the various answers are reported back to the main processor upon completion of each assigned task. Even *autochronous* systems have been suggested where a central processor, using a local clock, controls the flow of the problem solution and the remote processors each have their own local clocks to prevent clock skew problems. *Communication is expensive* and *computation is not,* so those systems that restrict the required amount of communication between the individual sections of a system will be most efficient.

Although improvements can be made in computer architectures, there is also much to be gained by improving the user-system interface. As computers are being used by an ever-increasing number of people who are not computer experts, low-cost systems that are easy to work with become very important. This tends to favor more human-oriented forms of communication with a computer such as speech (both recognition and synthesis), handwriting recognition, and even computer "vision."

8.2 THE SILICON FOUNDRY

The results of the relocation of the IC design responsibility to the system houses has forced the IC suppliers into a new way of doing business. This is called the *Silicon Foundry*. This name emphasizes the similarity between the new role of the IC supplier and the metal casting industry. Customers are free to take their tooling (*Customer-Owned Tooling*, COT) to any of the *shops* that can economically produce their required circuits.

Many observers are concerning themselves with the problem of processing compatibility between the various silicon foundries. The processing of silicon continues to be very dynamic and changes are still being made on the production lines. For this reason, it is still a little early to enter the foundry phase because these processing changes can cause problems with the foundry users. Actually, some breakpoints are becoming evident in the processing of silicon. For example, 3 μm feature size can be handled by a single-step, complete wafer masking. In contrast, 2 μm features and smaller require wafer steppers to solve the wafer size distortion problem. This therefore indicates that a 3 μm process could be standardized within the industry and this could then be used for circuits that do not represent the ultimate in complexity. Unfortunately, leading-edge ASICs will always demand the latest processing technology to guarantee that the resulting systems are the most competetive in the marketplace. Thus, 2-μm and smaller processes are replacing the older 3- to-5-μm processes for ASIC fabrication.

It may appear that computer-assisted design could also provide the ability to redo an existing mask set to conform to a new process. Unfortunately, this can be very difficult. The required effort may be close to that of a complete redesign of the IC.

The success of the silicon foundry concept depends on the easy portability of the tooling between at least two IC suppliers. This should not be a business problem, because, as any technology matures, the higher development and capitalization costs have a tendency to make the surviving suppliers cooperate more than they may have during the past years.

The very dynamic upheavals and changes that have always been characteristic of the IC manufacturing lines has tended to provide little time to obtain smooth-running, well-characterized wafer fabrication. Large investments are now being made that focus on "manufacturing engineering" and "manufacturing science" to provide a much more sophisticated "manufacturing technology." High degrees of automation are expected to reduce the variability of wafer fabrication processes (and will also reduce the particulate contamination problems). The needed change is to move away from the old way that required "firefighting" poorly controllable processes to achieve comparatively well-behaved, ultra-clean, stable and predictable wafer fabrication. In addition, there is also interest in coming up with three-dimensional computer models for each process used—including the interactions between the process steps. Test runs of wafers will then no longer be needed because computer simulation will be able to predict and isolate problems with a newly proposed process flow. Computers now help circuit, system, and mask designers. With the always-increasing complexity of wafer fabrication, it is not surprising that computers would also be beneficial to the process engineers.

8.3 SOFTWARE IN THE VLSI ERA

Software is very important in the new VLSI era. In most cases, a customer *sees* and really *interfaces* with software. The actual hardware that is used tends to be hidden, provided that it meets some basic performance goals. This shifting emphasis onto the importance of software is already evident in the software tools that are now necessary to do the design work. In the past, software has been a rather weak area for the IC suppliers but there is currently much support being added to this very important area.

8.4 THE LIMITS OF DEVICE SCALING

As lithography and scaling continue to reduce the feature sizes, the question "What is the fundamental limit on channel length in an MOS transistor?" is often asked. The success of the semiconductor industry has been based on schemes that have permitted the physical down-scaling of transistor-based ICs. As an indication of what can be expected in the future, experimental logic circuits have been reported by IBM researchers that were based on a 0.1-μm technology where the field effect transistors had features only a few hundred atoms wide! Researchers are also now considering other alternatives to transistor-based ICs.

These next generations of ICs must solve the interconnection problem. More complex functions must be performed within the devices (more than one device type is expected) than simply switching. These new devices are not expected to be transistors and will most likely be based on new phenomena, although transistors will still be used for some switching functions and interfacing. For example, quantum semiconductor devices have been reported that make use of charge carrier propagation by quantum mechanical tunneling within small structures where the charge carriers are confined by energy levels instead of the depletion layers that exist with conventional transistors. Only two-terminal devices have been reported, and researchers are attempting to implement a quantum device technology based on these devices.

8.5 GALLIUM ARSENIDE ICs

Gallium arsenide (GaAs) has rapidly moved away from only providing light-emitting diodes, and it now becomes clear that it is also important for high-speed and high-frequency circuit applications. Researchers have produced a 1K-bit static RAM, a 3K-gate array with gate delays under 100 ps, high-speed logic circuits, and microwave amplifiers.

GaAs circuits can be optimized for low power, rather than high speed, and they can also exhibit ten to one hundred times higher tolerance to radiation than the best silicon circuits. These circuits can still meet subnanosecond speeds; they are expected to be twice as fast as the best bipolar circuits. A several thousand gate complexity product can be expected at power levels below one watt. In addition, GaAs products can also operate at very high ambient temperatures.

The progress of GaAs ICs has been impeded because of poor wafer quality and also because this compound semiconductor will decompose at the elevated processing temperatures that are used for the single element semiconductors. New breakthroughs are being made by the use of ion implantation for the entry of the dopants. Current research is also transferring all of the knowledge gained in silicon processing over to this new semiconductor material. Therefore, silicon is being used as a dopant and deposited SiO_2 films are used as dielectric layers.

GaAs is not expected to become the future building block for large digital systems or for volume digital system components, such as memory, but it is expected to have application in special-purpose medium-scale digital and microwave circuits. Because of rapid changes in technology, it is worth keeping an eye on the progress in GaAs technology and products.

BIBLIOGRAPHY

1. Barbe, D. F.: *Very Large Scale Integration—VLSI Fundamentals and Applications*, Springer-Verlag, Berlin, West Germany, 1980.

2. Bryant, R. (ed.): *Third Caltech Conference on Very Large Scale Integration*, Computer Science Press, Rockville, MD, 1983.

3. Gibbons, T. J. and K. F. Lee: *One-Gate-Wide MOS Inverter on Laser-Recrystallized Polysilicon*, IEEE Electron Devices Letters, EDL-1, pp. 117–118, June, 1980.

4. Keyes, R. W.: *The Evolution of Digital Electronics Toward VLSI*, IEEE Trans. ED, vol ED-26, pp. 271–279, April, 1979.

5. Mead, C. and L. Conway: *Introduction to VLSI Systems*, Addison-Wesley, Reading, MA, 1980.

6. Mc Greivy, D. J. and K. A. Pickar: *VLSI Technologies Through the 80s and Beyond*, IEEE Computer Society Press, Silver City, MD, 1982.

7. Newton, R.: *Computer-Aided Design of VLSI Circuits*, Proc. IEEE, vol. 69, No. 10, pp. 1189–1199, October, 1981.

8. Sano, E., K. Ohwada and T. Kumura: *A Buried Channel/Surface Channel CMOS IC Isolated by an Implanted Silicon Dioxide Layer*, IEEE Trans. ED-29, pp. 459–461, March, 1982.

9. Monticelli, D. M.: *A Quad CMOS Single-Supply Op Amp with Rail-to-Rail Output Swing*, IEEE Journal of Solid-State Circuits, vol. SC-21, no. 6, pp. 1026–1034, December 1986.

10. Vyne, R. L., W. F. Davis, and D. M. Susak: *A Monolithic P-Channel JFET Quad Op Amp with In-Package Trim and Enhanced Gain-Bandwidth Product*, IEEE Journal of Solid-State Circuits, vol. SC-22, no. 6, December, 1987.

Index

About the Author

Thomas M. Frederiksen, engineer, author, and seminar leader, founded his own company, Intuitive IC Seminars, to provide instruction to electronic design engineers and the many nonelectronically educated people who either work for, or with, a high-technology electronic company.

Upon earning his BSEE degree from California Polytechnic State University at San Luis Obispo, Frederiksen started his professional career as a development engineer with the Motorola Systems Development Laboratory. Subsequently he worked with the Microelectronics Group at Hughes Semiconductor Division and later became a senior project engineer at Motorola Semiconductor Products Division. Mr. Frederiksen then joined National Semiconductor Corporation, Santa Clara, California, where he developed custom ICs and standard single-supply building-block circuits for automotive and industrial applications. He designed the popular Quads: LM3900, LM324, and LM339. The LM324 is today's most popular operational amplifier. He has also been involved with analog-to-digital converters that will interface with microprocessors and other data acquisition circuits.

Mr. Frederiksen is the author of five books: *Intuitive IC Electronics, Intuitive CMOS Electronics,* and *Intuitive Operational Amplifiers* for engineers and technicians and *Intuitive Digital Computer Basics* and *Intuitive Analog Electronics,* which introduce the basics of electronics and also digital and analog circuits and systems, for nonelectronic professionals.

Mr. Frederiksen holds more than 40 patents on linear ICs and devices, has been a frequent contributor to the professional literature, and has given many major seminars on linear ICs for both Motorola and National Semiconductor within the United States and Canada and abroad. In 1977 he received the International Solid State Circuits Conference Best Paper Award.